DAVID LINK

Archaeology of Algorithmic Artefacts

Archaeology of Algorithmic Artefacts
by David Link

First Edition
Minneapolis © 2016, Univocal Publishing

Published by Univocal Publishing
411 Washington Ave. N., STE 10
Minneapolis, MN 55401
www.univocalpublishing.com

Designed & Printed by Jason Wagner
Distributed by the University of Minnesota Press

ISBN 9781937561048
Library of Congress Control Number: 2016936891

Table of Contents

ARCHAEOLOGY OF ALGORITHMIC ARTEFACTS

Introduction

The more historical processes become steeped in technology, the more a discipline that is capable of comprehending these materialities becomes necessary. If scientists investigate the world with complex instruments, artists integrate advanced tools into their creations, and wars are no longer won by way of an arms race, but by way of a technological race, then without an examination of the underlying artefacts, no understanding of these activities is possible. This unfavorable situation is exacerbated when the apparatuses employed are algorithmic, for example, in the case of scientific, artistic, or militant calculations performed on computers. Because of their high complexity and complete arbitrariness, the technologies and languages used become unintelligible and are lost at an extremely fast pace.

Necessarily proceeding theoretically and practically at the same time, an Archaeology of Algorithmic Artefacts confronts this tendency of burying. It reconstructs the apparatuses in question and makes them run again in order to subject them to an effective analysis, but at the same time, because of the objects' complexity it can only perform this restoration with recourse to theoretical knowledge.

An Archaeology of Algorithmic Artefacts is devoted to manifold technologies, but there is one which plays a central role – the Turing Machine, or the computer. In an attempt to establish the ontology and genealogy of this special apparatus, a closer description of its construction is required, which draws us into the history of its origin, because in and of itself, the Turing Machine represents an artefact without properties. Its universality – the ability to simulate arbitrary discrete (and discretisable) apparatus – also leads to an absence of characteristics, because any concrete determination would prevent such generality.

This text investigates the gradual developments of the individual components encompassed by this complex technology within the context of the engineering sciences and the history of inventions. The material associations of artefacts –

which are only slowly changed by inventors and engineers through the processes of metaphor and metonymy, through the exchange and the tuning of parts – constitute genealogies of technical objects, which often last through the centuries in various functions and offer a stable fundament for their theoretical investigation.

This genealogy also consists in tracing the origin of the computer in other disciplines like mathematics, meta-mathematics, combinatorics, cryptology, philosophy and physics. The history of apparatuses that process signs is in no way limited, as one might think, to the second half of the twentieth century. It is possible that they existed at all times and in all cultures.

The first chapter compares the concept of a Universal Machine by the British mathematician Alan Turing with universal systems, which the German idealistic philosopher Georg W.F. Hegel developed approximately 130 years earlier. Regardless of there being no direct link and several differences between the two thinkers, there are nevertheless elements in each of their systems that help elucidate the other's thought. For example, the first concepts in Hegel's system, Being and Nothing, are completely general and precede the differentiation in letter and number, like the signs that the Universal Machine, and hence every computer, processes.

The second chapter is devoted to a direct predecessor of the Turing machine, Markov chains. In his seminal text of 1913, the Russian mathematician Andrei A. Markov discovered through counting letters in Pushkin's novel *Eugene Onegin* that the probability of a sign being a consonant depends on what precedes it – vowel or consonant. The chapter analyses his mathematical approach in detail and relates it to the cryptographic methods of his time. The technique can also be used to estimate how closely a text represents spoken language. Around the same time period, passages are found in Ferdinand de Saussure's *Course in General Linguistics*, which can shed light on Markov's purely mathematical explanations.

The third chapter focuses on the first computer realised worldwide, the Manchester Mark 1. The engineers developed the volatile storage that made this machine possible out of the so-called "Moving Target Indication" in the radar technology of the Second World War. The permanent increase in the range of the equipment finally led to hallucinations of the apparatuses themselves, which were discussed under the title of "radar angels" – dots on the CRT screen, which no longer corresponded to enemy aircrafts and which directly preceded the storage spots of the Mark I's memory – the Williams tube. In 1952, the British mathematician Christopher Strachey wrote a programme for the Mark I, which of all things composed love-letters. This chapter reconstructs Strachey's algorithm and offers a detailed analysis.

The fourth chapter develops general thoughts on the necessity and the methodology of a new discipline – an Archaeology of Algorithmic Artefacts. Serving as an example and emblem of this investigation is a strange creature from Kafka's short story *The House Father's Concern* – Odradek. It symbolises artefacts in the peculiar state they attain before and after they are in common use. Apart from that, objects are found in the history of technology over and over again, which show a close similarity to Odradek and may have inspired Kafka. On the basis of the reconstruction of the prototype of the Manchester Mark I, the Small Scale Experimental Machine (SSEM), we are able to identify which documents are absolutely necessary to restore an artefact that has gone extinct. This is followed by a close reading of Michael Thompson's "rubbish theory" wherein a theory of the cyclical return of technical objects is presented, in which these are communicated at every step of their life cycle.

In chapter 5, an Archaeology of Algorithmic Artefacts is applied to a central apparatus of the history of cryptology of the Second World War, the *bomba kryptologiczna*. Already in 1938, Polish mathematicians had broken the German cipher machine *Enigma* and built the *bomba kryptologiczna* to speed up the decryption process. Astonishingly, the descriptions in the literature of the methods used by this apparatus do not work when tried out on a computer simulation. By using an emulator and with recourse to the original reports written by the chief-cryptologist Marian Rejewski, this text reconstructs the actual deciphering technique used in the *bomba kryptologiczna* and decrypts an authentic dispatch with it.

By reaching far back into history, chapter 6 applies the new discipline to an approximately 800-year-old algorithmic artefact, the Maghrebian oracle device *zairja*. The Majorcan philosopher Ramon Llull imported this technology from the Muslims he was seeking to convert into the realm of the vowel alphabet. He would later found Western logic based on the imported technology of the *zairja*. From today's perspective, this device offered a highly astonishing function: For every question posed, it generated an answer that was not only truthful, but even rhymed. This was achieved by following a very complicated, but rule-based, procedure. This text reconstructs large parts of the complex algorithm for the first time and situates it within the history of the technological form of letter wheels that spans several centuries.

Chapter 7 returns to the early history of the first computer, the Manchester Mark I. In 1951, Christopher Strachey wrote a programme that mastered the game of draughts. After presenting the general structure of the software, the text provides a case study into the reconstruction of the algorithm. The programme seems to wait for something impossible, when examined on the purely symbolic level. But if the implementation of symbolic processes through corresponding components and the timing of the cycles of the machine are taken into consideration, it becomes apparent that this was how Strachey realised the function that is now achieved by pressing the "Enter"-key. Such a mechanism was missing on the first computer and on many of the succeeding ones. The chapter concludes with the history of the gradual development of the computer "keyboard" from out of teleprinters and of the origins of the first "Enter"-key.

Alan Turing is often considered as a slightly unworldly, a pure mathematician, compared to the engineers that actually constructed the computers. The last chapter shows the contrary: from early on and throughout his life, Turing built calculating devices or was directly involved with their realisation. After constructing a number of specialised apparatuses, in 1948, his dream from 1936 of a Universal Machine becomes a technical reality and made programming possible. The second part of the text presents the algorithms that Turing wrote on the Manchester Mark I or in whose creation he was significantly involved.

while(true)
On the Fluidity of Signs in Hegel, Gödel, and Turing

The universal machine, which the Englishman Alan Turing designed and later constructed, exhibits in principle a number of structural similarities to systems that the German philosopher Georg Wilhelm Friedrich Hegel began to develop 130 years earlier. By stating that their logical constructions manifest a considerable degree of closeness, I am not suggesting that Hegel anticipated the computer. Nor is there any evidence that Turing sought to imitate Hegel's system of thought. Indeed, during Turing's student days in 1930s Cambridge the intellectual climate was decidedly anti-Hegelian. In 1914, the analytical philosopher Bertrand Russell had published a refutation of Hegel, *Our Knowledge of the External World,* in which he attempts to show that the idealist misapprehends the meaning of the copula "is."[1] However, the proximity of Hegel's and Turing's systems makes it possible to locate them in a history of the mechanisation of thought as a means of understanding and to illuminate the one from the perspective of the other.

0 !＝ 1

Hegel's *Science of Logic* begins with two concepts: "being" and "nothing." Both concepts are pure symbols in the sense that they mean nothing, that is, they do not reference anything particular in the external world, but merely point to the void: "[I]t is altogether the same as what the Indian calls Brahma, when for years on end, physically motionless and equally unmoved in sensation, conception, phantasy, desire and so on, looking only at the tip of his nose, he says inwardly only *Om, Om, Om,* or else nothing at all."[2] These symbols are so

---

1. Bertrand Russell, *Our Knowledge of the External World* (Chicago and London, 1914).

2. "OMOMOM" looks digital and can be interpreted as "42" ("101010"); see Douglas Adams, *The Hitchhiker's Guide to the Galaxy* (London, 1979).

general that they even precede the differentiation into letters and numerals and can be named at will: "With this [...] indeterminateness and vacuity of conception, it is indifferent whether this abstraction is called space, pure intuiting, or pure thinking."[3] This pre-literacy can also be recognised from the circumstance that the nothingness of being also emerges when one does not speak of it or write it down, but simply when it is shown, as at the beginning of Hegel's *Phenomenology of Spirit:* the Now "has already ceased to be in the act of being pointing to it."[4] Decisive for further progress is only that the signs are distinct and thereby mark the difference between them.

The first two symbols of Hegel's system no longer reference the external world, also because the adequation theory of truth had failed, at the latest in Immanuel Kant's theory of the "thing in itself."[5] Instead of comparing words with things, as adequation theory does, Hegel chooses an approach that is purely internal to consciousness and only compares concepts: "Consciousness provides its own criterion from within itself, so that the investigation becomes a comparison of consciousness with itself."[6] The second reason for the emptiness of the symbols is that they are at the beginning of Hegel's system and, therefore, have to be undetermined and unmediated. Any specific content would contradict this because, being mediated, something would precede it.[7]

During Hegel's lifetime, the way was cleared for a further breakaway of the symbolic from the real world. In 1829, the Russian mathematician Nikolai Ivanovich Lobachevsky publishes an essay on hyperbolic geometry, which later results in a fundamental crisis in mathematics.[8] Interest in non-Euclidean spaces increases through the publications of Bernhard Riemann and Felix Klein, the latter of which went on to marry one of Hegel's granddaughters in 1875.[9] Simple intuitive basic assumptions, such as Euclid's fifth axiom, which states that non-parallel straight lines extended indefinitely cross just once,

3. Georg Wilhelm Friedrich Hegel, *Science of Logic* [1812–1831], trans. A.V. Miller (London, NJ, 1990), p. 97.

4. G.W.F. Hegel, *Phenomenology of Spirit* [1807], trans. A.V. Miller (Oxford, 1979), p. 63.

5. Cf. Immanuel Kant, *Critique of Pure Reason* [1781–1787], trans. P. Guyer (Cambridge, 1998), p. 185, B 59: "[W]hat may be the case with objects in themselves and abstracted from all this receptivity of our sensibility remains entirely unknown to us."

6. Hegel, *Phenomenology*, p. 53.

7. Cf. Hegel, *Logic*, p. 67f.: "With What Must the Science Begin?"

8. Nikolai Ivanovich Lobachevsky, The fundaments of geometry. *Kazanskii Vestnik* 25, 27, and 28 (1829), in Russian. Both sources and content of this paper are very unclear. An alternative first publication is: Janos Bolyai, Appendix, scientiam spatii absolute veram exhibens a veritate aut falsitate Axiomatis XI Euclidei (a priori haud unquam decidenda) independentem: adjecta ad casum falsitatis, quadratura circuli geometrica, in: Farkas Bolyai, *Tentamen in Elementa Matheseos Purae, Elementaris ac Sublimioris, Methodo Intuitiva, Evidentiaque huic propria, Introducendi. Cum Appendice Triplici* (Marosvásárhely, 1832).

9. Bernhard Riemann, *Ueber die Hypothesen, welche der Geometrie zu Grunde liegen* [1854]. *Abhandlungen der Königlichen Gesellschaft der Wissenschaften zu Göttingen* 13 (1868); Felix Klein, Über die sogenannte nicht-euklidische Geometrie. *Nachrichten von der Königl. Gesellschaft der Wissenschaften und der Georg-Augusts-Universität zu Göttingen* 17 (1871): 419–433.

are demonstrated to be false when applied to spherical surfaces.[10] In 1895, Felix Klein added David Hilbert from Königsberg to his research team in Göttingen. According to Hilbert, it is necessary to abandon all reference to the real world through counting and measuring and, instead, establish geometry as an abstract system of symbols that does without intuitive and illustrative assumptions with the aim of giving metamathematical proofs of its consistency: "one must always be able to say, instead of 'points, straight lines, and planes,' 'tables, chairs, and beer mugs.'"[11] The mathematicians decide to uncouple their discourse completely from external reality and ground the world of numbers entirely within itself. Although Hilbert's words do not appear to forswear the world of the senses, this step is based upon the distrust of sensory perception, which is seen as deceptive, a widespread view that has prevailed since classical antiquity.[12] From this point in time, numbers are treated as "any system of things,"[13] without reference to the world and ignoring their ordinal nature. Being formalistic, mathematics is a game with empty symbols, which offer as little to apprehend or contemplate as "being."

Similarly, Turing's concept of a universal machine, as described in his essay "On Computable Numbers," is part of this tradition and utilises symbols that precede the distinction between letters and numbers, reference nothing, and are completely meaningless. The essay text is itself a Babylonian mixture of indifferent signs. In addition to Arabic numbers and Roman capital and small letters, Turing uses Fraktur capitals and small letters, symbols from predicate calculus (∂), Roman capital letters in cursive writing, and Greek capitals and small letters.[14] In the descriptive language of the essay, these symbols serve to differentiate between various abstract entities ("classes"). In the symbol set of the machine he describes, like "being" and "nothing," they mark the pure difference, to which meaning can only be ascribed subsequently and arbitrarily. If the machine's tape is inscribed with OMOMOM and the programme transforms this into OOMOOOMMOMM, at first it cannot be decided whether the machine has squared 42 or composed a new mantra. Turing machines can thus be constructed from almost any kind of material, such as DNA, mirrors, model railways, or

---

10. Cf. Euclid, *Elements* [ca. 300 BC], trans. I. Todhunter (London, 1933), p. 6: "[I]f a straight line meets two straight lines, so as to make the two interior angles on the same side of it taken together less than two right angles, these straight lines, being continually produced, shall at length meet on that side on which are the angles which are less than the two right angles."

11. Otto Blumenthal, David Hilberts Lebensgeschichte [1935], in: David Hilbert, *Gesammelte Abhandlungen*, vol. 3 (New York, 1965), pp. 388–429, here p. 403. Hilbert is said to have made this remark in 1891 on the way home from a lecture by Hermann Wiener.

12. Cf. Plato, *Phaedo* [ca. 387 BC], trans. R. Hackforth (London, 1972), p. 83: "Now were we not saying some time ago that when the soul makes use of the body to investigate something […] it is dragged by the body towards objects that are never constant, and itself wanders in a sort of dizzy drunken confusion, inasmuch as it is apprehending confused objects?"

13. D. Hilbert, Problems of the grounding of mathematics [1928], in: Paolo Mancosu, ed., *From Brouwer to Hilbert: The Debate on the Foundations of Mathematics in the 1920s* (New York, 1998), pp. 227–233, here p. 232.

14. Alan M. Turing, On computable numbers, with an application to the Entscheidungsproblem. *Proceedings of the London Mathematical Society (Ser. 2)* 42 (1937): 230–265.

hosepipes.[15] The Manchester "Baby" computer, which was completed in 1948 with the participation of Alan Turing, worked with a base 32 alphabet (5-bit). Its first symbol was neither a letter nor a number but a forwardslash, /. "The result was that pages of programmes were covered with strokes – an effect which at Cambridge was said to reflect the Manchester rain lashing at the windows."[16] Its memory and processor were cathode ray tubes, that is, television screens; thus the workings and outputs of the machine were seen as a "mad dance" of flickering dots on the screen, which was a "dance of death" until the final breakthrough on 21 June 1948.[17] The fact that the symbols are not letters or numbers particularly endows the ideas of Turing and Gödel with their fundamental power. Their ideas do not concern specific symbols but symbols in general. This also avoids assigning to the computer a one-sided bias: either to the field of letters or to the field of numbers. A Turing machine only processes that which stands at the beginning of Hegel's *Logic*: pure difference. The only condition for the symbols is that they are distinguishable; that is why their number is finite.[18]

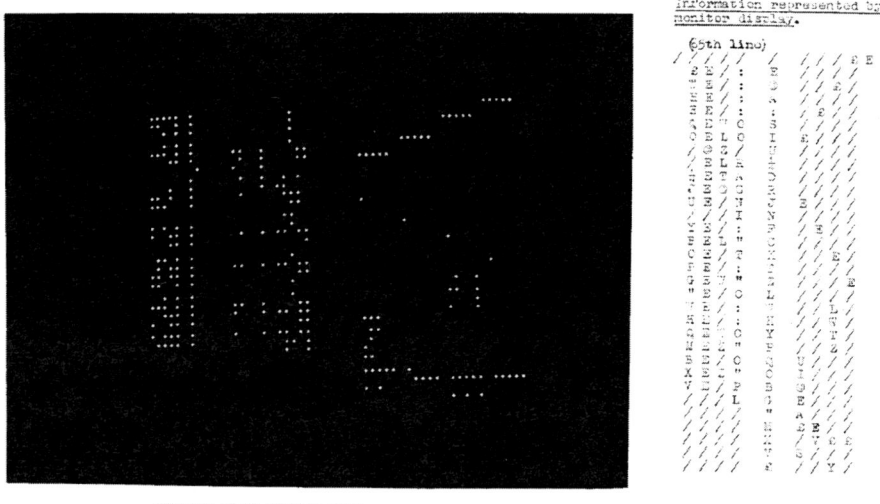

Fig. 1: The screen of the Manchester computer and the information represented.

15. See Y. Benenson, T. Paz-Elizur, R. Adar, E. Keinan, Z. Livneh, and E. Shapiro, Programmable and autonomous computing machine made of biomolecules. *Nature* 414 (2001): 430–434; Christopher Moore, Predictability and undecidability in dynamical systems. *Physical Review Letters* 64 (1990): 2354–2357; Severin Hofmann and David Moises, Turing Train Terminal, 2003/4; online: http://www.monochrom.at/turingtrainterminal/pictures_eng.htm; Paulo Blikstein, Programmable Water, 2003; online: http://www.blikstein.com/paulo/projects/project_water.html.

16. Andrew Hodges, *Alan Turing: The Enigma* (New York, 1983), p. 399.

17. Ibid., p. 392; cf. Figure 1.

18. Turing, On computable numbers, p. 249: "If we were to allow an infinity of symbols, then there would be symbols differing to an arbitrarily small extent."

0 = 1

Because the sign "being" has no content, it passes over into "nothing." The difference between the two is merely supposed. They are not different but should only be distinguished. "Nothing" is an empty symbol and, therefore, the same as "being." Both symbols are abstract and meaningless: "*Pure being* and *pure nothing* are, therefore, the same."[19] Hegel writes that this "thesis" is "so paradoxical," "indeed [...] one of the hardest tasks thought poses for itself."[20] Throughout the three main works of Hegel, there are only two instances of the word "paradox," or "paradoxical." In paragraph 104 of the *Encyclopaedia*, Pythagoras' basic determination of things as numbers is referred to as such, and in connection with this, in paragraph 301, phenomena of the objective appearance of harmony, for example, that one string can produce several notes, several strings just one note, or two strings a third note, and so on. Paradoxically, Hegel uses the word very seldom because it is fundamental to his theory and is throughout denoted by the concept of "dialectic." Hegel points out that, in general, we distinguish between things on the basis of some common ground, for example, between two species of the same genus: "In contrast, with Being and Nothing the difference is in its bottomlessness and, therefore, is none, since both determinations are the same bottomlessness."[21] The difference between the two symbols breaks down because they are both empty and, therefore, the same. At the same time, the difference continues to exist and for this reason the sentence is paradoxical, as are all assertions of identity. The sentence distinguishes between the two symbols and relates them to each other at the same time. It represents the identity of identity and non-identity, and is paradigmatic of speculative thought: "So, too, in the philosophical proposition the identification of Subject and Predicate is not meant to destroy the difference between them, which the form of the proposition expresses; their unity, rather, is meant to emerge as a harmony."[22] The sentence does not consist in the fact that attributes are ascribed to a subject, but that the signs change into each other in the "movement of the Notion," like two vibrating strings that produce a third note.[23] The "fixed thoughts" are transformed into "a fluid state" and set in motion.[24]

---

19. Hegel, *Logic*, p. 82.

20. G.W.F. Hegel, *Encyclopedia of the Philosophical Sciences in Outline and Critical Writings* [1830], trans. E. Behler (New York, 1990), vol. 1, p. 70.

21. "Beim Sein und Nichts dagegen ist der Unterschied in seiner Bodenlosigkeit, und eben darum ist es keiner, denn beide Bestimmungen sind dieselbe Bodenlosigkeit." Translation, D.L. (G.W.F. Hegel. *Enzyklopädie der philosophischen Wissenschaften. Werke 8, 9, 10* [1830] (Frankfurt a. M., 1970), vol. 8, p. 187).

22. Hegel, *Phenomenology*, p. 38.

23. "It is the same with the other case, where, when following Tartini two different strings of a guitar are strummed, the wonderful happens, that apart from their sound a third sound is heard that, however, is not a mere mixing of the first two, not only an abstract neutral." ("Ebenso ist es dann auch mit dem anderen Fall, wo, wenn man nach Tartini, zwei verschiedene Saiten einer Gitarre anschlägt, das Wunderbare geschieht, daß man außer ihren Tönen auch noch einen dritten Ton hört, der aber nicht bloß die Vermischung der beiden ersten, kein bloß abstrakt Neutrales ist.") Translation, D.L. (Hegel, *Enzyklopädie*, vol. 9, p. 183).

24. Hegel, *Phenomenology*, p. 20.

In 1899, Hilbert proved the consistency of Euclidean geometry under the supposition that the theory of real numbers is consistent. In 1872, Richard Dedekind succeeded in deriving real numbers from whole numbers. The question thus became, is the theory of whole numbers consistent?[25] At the International Congress of Mathematicians in 1900 in Paris, Hilbert presented a list of tasks for the coming century in the form of 23 unsolved mathematical problems. Problem No. 2 concerned the freedom from contradiction of the arithmetical axioms.[26] Gottlob Frege attempted to solve this with a further development of Cantor's set theory, which avoids reference to the ordinal character of numbers. However, in 1901, the 30-year-old Bertrand Russell came upon the very paradoxical concept of the set of all sets, which do not contain themselves. This set contains itself when it does not contain itself and vice versa. In contrast to the Cretan paradox, it is conclusive.[27] In a letter dated 16 June 1902, Russell communicated his discovery to the 54-year-old Frege, who was just preparing the second edition of his book, *Grundsätze der Arithmetik*.[28] Frege's reply was one of humility – and astonishment: "Your discovery of the contradiction has surprised me beyond words and, I should almost like to say, left me thunderstruck, because it has rocked the ground on which I meant to build arithmetic."[29] Frege added a resigned afterword to the second volume of his book and, after this incident, gave up set theory.

In Hegel, thought was driven to "nothing," when it attempted to hold fast to "being," and to "being," when it attempted to hold fast to "nothing," moving eternally in circles. Here again, thought finds itself entering into a vertiginous circular motion that Russell reflects upon in the following manner: "giving a person a piece of paper on which is written: 'The statement on the other side of this paper is false.' The person turns the paper over, and finds on the other side: 'The statement on the other side of this paper is true.'" Russell was convinced that the origin of such paradoxes lay in the self-application of statements, and over the next ten years he developed a theory of types, together with his teacher, Alfred North Whitehead, which attempted to prevent such mixing – the *Principia Mathematica*. The profundity, lengthiness, and difficulty of their undertaking can be gauged by the fact that $1 + 1 = 2$ is only proved on page 379.[30]

---

25. D. Hilbert, *Grundlagen der Geometrie* [1899] (Stuttgart, 1968); Richard Dedekind, *Stetigkeit und irrationale Zahlen* (Braunschweig, 1872).

26. D. Hilbert, Mathematical problems. *Bulletin of the American Mathematical Society* 8 (1902): 437–479, here p. 447: "But above all I wish to designate the following as the most important among the numerous questions which can be asked with regard to the [arithmetical, D.L.] axioms: *To prove that they are not contradictory, that is, that a finite number of logical steps based upon them can never lead to contradictory results.*"

27. In its usual form, the Cretan paradox is not conclusive because the opposite of "all Cretans lie" in predicate logic is "one Cretan tells the truth." Thus the Cretan who speaks is lying and some other Cretan tells the truth.

28. Cf. Jean van Heijenoort, *From Frege to Gödel. A Source Book in Mathematics, 1879–1931* (Harvard, 1967), p. 124f.; Hodges, *Enigma*, p. 84.

29. Heijenoort, *Source Book*, p. 127f.

30. Alfred North Whitehead and Bertrand Russell, *Principia Mathematica*, 3 vols (Cambridge, 1910, 1912, 1913), vol. 1, p. 379; cf. Figure 2.

*54·42. $\vdash :: \alpha \,\epsilon\, 2 . \supset :. \beta \subset \alpha . \exists ! \beta . \beta \neq \alpha . \equiv . \beta \,\epsilon\, \iota``\alpha$

  Dem.

  $\vdash . *54·4 . \supset \vdash :: \alpha = \iota`x \cup \iota`y . \supset :.$

  $\qquad\qquad \beta \subset \alpha . \exists ! \beta . \equiv : \beta = \Lambda . \mathsf{v} . \beta = \iota`x . \mathsf{v} . \beta = \iota`y . \mathsf{v} . \beta = \alpha : \exists ! \beta :$

  [*24·53·56.*51·161] $\equiv : \beta = \iota`x . \mathsf{v} . \beta = \iota`y . \mathsf{v} . \beta = \alpha$ (1)

  $\vdash . *54·25 . \text{Transp} . *52·22 . \supset \vdash : x \neq y . \supset . \iota`x \cup \iota`y \neq \iota`x . \iota`x \cup \iota`y \neq \iota`y :$

  [*13·12] $\supset \vdash : \alpha = \iota`x \cup \iota`y . x \neq y . \supset . \alpha \neq \iota`x . \alpha \neq \iota`y$ (2)

  $\vdash . (1) . (2) . \supset \vdash :: \alpha = \iota`x \cup \iota`y . x \neq y . \supset :.$

  $\qquad\qquad \beta \subset \alpha . \exists ! \beta . \beta \neq \alpha . \equiv : \beta = \iota`x . \mathsf{v} . \beta = \iota`y :$

  [*51·235] $\equiv : (\exists z) . z \,\epsilon\, \alpha . \beta = \iota`z :$

  [*37·6] $\equiv : \beta \,\epsilon\, \iota``\alpha$ (3)

  $\vdash . (3) . *11·11·35 . *54·101 . \supset \vdash . \text{Prop}$

*54·43. $\vdash :. \alpha, \beta \,\epsilon\, 1 . \supset : \alpha \cap \beta = \Lambda . \equiv . \alpha \cup \beta \,\epsilon\, 2$

  Dem.

  $\vdash . *54·26 . \supset \vdash :. \alpha = \iota`x . \beta = \iota`y . \supset : \alpha \cup \beta \,\epsilon\, 2 . \equiv . x \neq y .$

  [*51·231] $\equiv . \iota`x \cap \iota`y = \Lambda .$

  [*13·12] $\equiv . \alpha \cap \beta = \Lambda$ (1)

  $\vdash . (1) . *11·11·35 . \supset$

  $\qquad \vdash :. (\exists x, y) . \alpha = \iota`x . \beta = \iota`y . \supset : \alpha \cup \beta \,\epsilon\, 2 . \equiv . \alpha \cap \beta = \Lambda$ (2)

  $\vdash . (2) . *11·54 . *52·1 . \supset \vdash . \text{Prop}$

From this proposition it will follow, when arithmetical addition has been defined, that $1 + 1 = 2$.

Fig. 2: The proof of 1 + 1 = 2 in Whitehead's and Russell's Principia Mathematica.

These flickering sentences, which continually drive thinking in circles are dangerous because they annul mathematics as a decision-making procedure between true and false and push it into a realm of absolute falsehood and absolute truth. Hodges writes of the problem of freedom from contradiction: "And that spelt disaster. In any purely logical system there was no room for a single inconsistency. If one could ever arrive at '2 + 2 = 5' then it would follow that '4 = 5' and '0 = 1', so that any number was equal to 0, and so that every proposition whatever was equivalent to '0 = 0' and therefore true."[31] The system would be supra-universal and would lose all possibility of distinguishing between true and false. To illustrate the term "supra-universal," I cite a statement by the US-American Defence Minister Donald Rumsfeld, who interprets the bestial treatment of Iraqi prisoners in Abu-Ghraib prison as a positive sign: "The system worked."[32]

---

31. Hodges, *Enigma*, p. 84; the emphasis is mine – D.L.

32. Donald Rumsfeld, NBC Interview, 5 May 2004; online: http://www.defense.gov/Transcripts/Transcript.aspx?TranscriptID=2986.

The crisis in mathematics came to a completion when, in 1931, Kurt Gödel proved that all formal systems, like the *Principia Mathematica*, must necessarily contain undecidable propositions and, in contrast to reason, cannot achieve certain true propositions. In 1928, Hilbert once again formulated his project in three questions: Is mathematics complete? Is it free of contradictions? Is it decidable? With Gödel's discovery, Hilbert's project had failed. Whereas Russell had still assumed that self-application from time to time generates paradoxes, Gödel demonstrated that paradoxes are a necessary part of all sufficiently complex formal systems and cannot be eliminated. Hegel formulates this in an even more radical fashion when he states, "that every actual thing involves a coexistence of opposed elements. Consequently, to know, or, in other words, to comprehend an object is equivalent to being conscious of it as a concrete unity of opposed determinations."[33] The fundament in reality that had been lost with the advent of non-Euclidean geometry cannot be recovered through formalistic attempts to ground mathematics.

The catastrophe of the paradox that it is not possible to distinguish between "being" and "nothing," "true" and "false," or "0" and "1" and that one passes into the other in circles, is resolved in a fruitful manner by Alan Turing. The infinite transformation of symbols, that have become meaningless, into one another is the *modus operandi* of the Turing machine. If the machine is in a state "0" when the symbol "0" is encountered and the command "1R1" is executed, this happens twice. Both in the machine's memory and on the paper tape "0" changes into "1." To program an addition, exactly one "0" (the one between the blocks of summands) must be transformed into "1" and one "1" into "0." Thus the chain of symbols "OOOMOO" (3 + 2) becomes "OOOOOM," a result that can be interpreted as 5. Instead of laying down rules of transformation for how one true statement can be derived from another, as in formalist mathematics, Turing constructs a general machine for transforming symbols. From this point onwards, the truth or falseness of the transformation lies in the hands of the software developer. For this reason, the very first programmers' manual, Turing's *Programmers' Handbook for Manchester Electronic Computer Mark II*, devotes a great deal of time to specifying how errors in the software can be detected and rectified.[34] The Turing machine implements the identity of identity and difference directly, namely, technically. Empty symbols, which exhibit a bottomless difference and pass over into each other, are its foundation.

---

33. "[…] alles Wirkliche entgegengesetzte Bestimmungen in sich enthält und daß somit das Erkennen und näher das Begreifen eines Gegenstandes eben nur so viel heißt, sich dessen als einer konkreten Einheit entgegengesetzter Bestimmungen bewußt zu werden." Translation, D.L. (Hegel, *Enzyklopädie*, vol. 8, p. 128).

34. A.M. Turing, *Programmers' Handbook for Manchester Electronic Computer Mark II* (Manchester, 1951; online: http://www.alanturing.net/turing_archive/archive/m/m01/M01-001.html), p. 59: "Programming is a skill best acquired by practice and example rather than from books. The remarks here are therefore quite inadequate." Notwithstanding, Turing proceeds to elaborate this theme on the following 24 pages.

"(" = 11

In formalist mathematics, not only are letters and numbers the same thing, namely symbols, but also the difference between numbers, operations with them, and statements about them vanishes in this generality. Gödel and Turing are able to answer Hilbert's questions in the negative because they make functions and numbers interchangeable and thus have the potential to encode metamathematical statements as numbers and apply these statements to themselves. Logical symbols, such as disjunction, negation, and generalisation are simply translated into natural numbers at the beginning of Gödel's essay:

$$„0" \ \ldots 1 \qquad „\vee" \ldots 7 \qquad „(" \ldots 11$$
$$„f" \ \ldots 3 \qquad „\Pi" \ldots 9 \qquad „)" \ldots 13$$
$$„\sim" \ldots 5$$

"Naturally, for metamathematical considerations, it makes no difference which objects one takes as primitive symbols, and we decide to use natural numbers for that purpose."[35] By means of a system of prime number exponents, as already used by Leibniz, Gödel prevents the collision of operators coded in numbers and real natural numbers.[36]

To put it simply, the Turing machine also consists of a chain of symbols, which represents the data computed – the symbols on the tape – and another chain, which specifies the operations – the programme. However, the particular dynamics and universality arises from the circumstance that the symbols on the tape also determine the way the programme runs. They are instructions in the form of markers and data at the same time. Only through this can a general routine be written that adds two numbers. Furthermore, each particular Turing machine as a whole is a chain of symbols that feeds into the universal machine.

Similarly, in Hegel the dialectic is first set in motion when the meaning of the empty symbol "being" is thematised by further symbols. It is only possible to state that "being" is "nothing" within a blend of meta-language and object-language. It is "natural," because "'[n]otion' and 'object' […] both fall *within* that knowledge which we are investigating."[37] Hegel depicts the inside of a self-critical mind, which makes a concept like "being" from the world and always returns to it in order to check whether the concept coincides with his meaning. Each time the realisation of a concept's limitations forces the mind to move on to a new

---

35. Kurt Gödel, On formally undecidable propositions of *Principia Mathematica* and related systems [1931], in: *The Undecidable. Basic Papers on Undecidable Propositions, Unsolvable Problems, and Computable Functions*, ed. Martin Davis (Hewlett, NY, 1965), pp. 4–38, here pp. 7 and 13.

36. On Leibniz, see Gerhard F. Strasser, *Lingua Universalis. Kryptologie und Theorie der Universalsprachen im 16. und 17. Jahrhundert* (Wiesbaden, 1988), p. 241.

37. Hegel, *Phenomenology*, p. 53.

concept, such as "becoming," which nevertheless contains the identity of "being" and "nothing" as its aspects and is reflected again. As in the approaches of Gödel and Turing, it is the permanent self-application of statements that produces paradoxes that do not allow the mind to stop turning.

while(true){ }

At the system's end, it bends back to its beginning and thus forms a cycle of endless becoming. The "absolute idea," the highest concept in Hegel's *Science of Logic* ultimately ends by transforming into "being": "[T]he science exhibits itself as a *circle* returning upon itself, the end being wound back into the beginning, the simple ground, by the mediation; this circle is moreover a *circle of circles*."[38] This is necessary in order to subsequently motivate the (at first) undeterminate and groundless beginning. A deductively progressing system with a claim to universality must, at its end, when it has deduced the totality of all facts, turn back to its beginning because its beginning is the only thing that has not yet been deduced. Also symbolic spaces, which achieve universality through recombination, like Jorge Luis Borges' *Library of Babel*, which is "total" and whose "shelves register all the possible combinations of the twenty-odd orthographical symbols [...]: in other words, all that it is given to express, in all languages," arrive at their end only to return to their begining: "*The Library is unlimited and cyclical. If an eternal traveller were to cross it in any direction, after centuries he would see that the same volumes were repeated in the same disorder.*"[39] The founding father of combinatory systems, Raimundus Lullus, used circular disks to set his text machine in motion.[40]

Turing's construction is also an endless iterative loop. Contrary to popular belief, the programmes proposed in *On Computable Numbers* never stop. The text distinguishes between "circular" and "circle-less" ones. The circular programmes specify the computed real number endlessly by giving it further digits through sub-routines. The circle-less programmes reach "a configuration from which there is no possible move," or run on but do not print out any further numerical symbols.[41] Algorithms that stop and deliver a result do not occur in the first design of the universal machine, only algorithms that fail or remain in a state of becoming, endlessly modifying the result.

The majority of today's computer programmes are also designed to run endlessly. To this end, the software developer encloses the core of the algorithm in a so-called run-loop, an iterative loop, which is executed under the tautological condition "while(true)" or "while(1)." It is precisely this basis, that true

---

38. Hegel, *Logic*, p. 842.

39. Jorge Luis Borges, *Labyrinths. Selected Stories and Other Writings* [1944], trans. D. Yates (Harmondsworth, 1979), pp. 81 and 85.

40. See Amador Vega, *Ramon Llull and the Secret of Life* (New York, 2003), p. 62f.

41. Turing, On computable numbers, p. 233.

remains true and 1=1, which was shattered by Gödel's developments. If we were to formalise the actual basis of software, it would be: while(0 != 1 & 0 == 1).

"while(true)" secures the algorithm against its own bottomlessness. It still reveals an echo of the shock triggered by Gödel.

stop()

Hegel distinguishes between two forms of the identity of identity and difference of "being" and "nothing," namely, "becoming" [*Werden*] and "determinate being" [*Dasein*]. In the former, "being" and "nothing" are only present in the form of vanishing into each other, origination and passing. Becoming, however, must "also vanish. Becoming is as if it were a fire, which dies out in itself, when it consumes its material." The further negation of "becoming" and the result of this process is "determinate being," definite being, and is thus different from other being – this is the side of negation.[42]

The first occurrence of "determinate being" in computer science appears in an article that is only three pages long, by the mathematician Emil Post. Post independently developed ideas strikingly similar to those of Turing at around the same time period. His basic model is not like Turing's – an idealised teleprinter – but instead, a production line worker, who processes a series of boxes according to instructions, which he can again mark with a forwardslash, /. Post distinguishes between three types of commands: the first is independent of any mark, the second is a case differentiation, which commands this or that depending on whether there is a forwardslash or not, and the third command is "stop."[43] For Post, a programme is only considered as the solution of a problem when the process, which the programme sets in motion, stops for each specific input, that is, it reaches the third command.

In a similar way, in Tibor Rado's much later reformulation of the Turing machine the property of circle-lessness is interpreted in such a way that the programmes reach the so-called stop sign and leave a certain result on the tape, whereas circular programmes modify the symbols in a state of eternal becoming.[44] The modification of Turing's approach, which continues to exert influence today, is indebted to Kleene's authoritative text *Introduction to Metamathematics* from 1952: "Our treatment here is closer in some respects to Post 1936."[45]

---

42. Hegel, *Logic*, p. 105f. "[S]o ist es [das Dasein, D.L.] selbst ein Verschwindendes, ein Feuer gleichsam, welches in sich selbst erlischt, indem es sein Material verzehrt." Translation, D.L. (Hegel, *Enzyklopädie*, vol. 8, p. 195).

43. Emil Post, Finite combinatory processes. Formulation I [1936], in: Davis, *Undecidable*, pp. 288–291, here p. 290.

44. Cf. Tibor Rado, On non-computable functions. *The Bell System Technical Journal* 41 (1962): 877–884, here p. 877f.: "The last column [...] contains the index of the next card to be used, or 0 (zero), where 0 is the code for 'Stop.'"

45. Stephen Cole Kleene, *Introduction to Metamathematics* (Amsterdam, 1952), p. 361.

"Becoming is an unstable unrest which settles into a stable result."[46] One difference between Hegel and Turing is that in Hegel's system, reflection upon the limitations of the empty abstract concepts increasingly fills them and they become more concrete. On the other hand, as far as the machine's tape is concerned – when it stops at all – there is only a chain of still empty symbols. The device can go back to what it has written, but it cannot reflect upon it, and therefore cannot move it to a higher level ["aufheben"]. Because the machine does not understand what it is doing, it can only achieve the level of simple "determinate being," provided it does not remain in the state of eternal becoming. Even though the heights of content of the absolute idea, which are attained in Hegel's reading and writing, remain closed to the apparatus, it is still the case, as Alan Turing wrote 50 years ago, that if "the brain work[s] in *some* definite way," it can be emulated by the universal machine.[47]

With the construction of his machine, Turing reveals the uncanny finding that the basis of mathematical operations, upon which people had hoped and expected to base the foundation of all sciences, lacks calculability in even the simplest operations, if it is not sought in reality but in the formal processing of any "system of things." There is no general procedure to determine whether a programme is circular or circle-less, whether it will ever manage to inscribe "1" on paper, or how high the maximum result will be. In Tibor Rado's "Busy Beaver game," where the goal is to programme an n-state computer to write as many symbols as possible on a tape and then stop, competitors must inform the jury how many operations the entered program will make before it shuts down; otherwise, the jury cannot judge the algorithm.[48]

exit(0)

To answer the universal claim of Hilbert's theory of proof, Turing must develop a system that is capable of formulating all possible data and operations. Otherwise, there is a danger that a further extension of the machine would in fact allow the *Entscheidungsproblem* to be solved. His counter-proof is influenced by universalism and outdoes it, in that although he demonstrates that there can be no general procedure to distinguish between true and false, he constructs a formalism that can express all true and false propositions and methods of contemporary and future systems of axioms. For this reason it is pointless to think beyond the Turing machine or, indeed, beyond the binary.

---

46. Hegel, *Logic*, p. 106.

47. Hodges, *Enigma*, p. 420. Here Turing was obviously not thinking of the simulation of neurons: It is possible, he wrote, "to copy the behaviour of nerves, but there seems very little point in doing so. It would be rather like putting a lot of work into cars which walked on legs instead of continuing to use wheels." (Hodges, *Enigma*, p. 404).

48. Rado, Non-computable functions, p. 879: "Beyond the enormous number of cases to survey, he will find that it is very hard to see whether certain entries do stop at all. This is the reason for the requirement that each contestant must submit the shift number s with his entry."

The various positions, for example, of Hegel's *Phenomenology of Spirit*, are also universal. With the unreflected concepts of "I," "this," "now," and "here," sensorial certainty has developed a complete understanding of the world, for thus everything is either "I" or "this."[49] This generality makes it possible to refute sensorial certainty using certain examples, such as writing down the "now" of the night that will have passed.[50]

In addition, Hegel's system is "supra-universal" in the sense that he develops the supposed universal truth of a position in such a way that its falsity and finiteness is demonstrated, which forces its negation in the form of a new truth. The concepts are in a state of perpetual transformation. They are not being or nothing, but becoming: an endless transition between true and false. Also Hegel's system claims to run through all possible, that is, also all future positions. The difficulty of answering this prophecy is reflected by Michel Foucault almost 170 years later, on the occasion of his inaugural lecture at the Collège de France:

> "[O]ur entire epoch, whether in logic or epistemology, whether in Marx or Nietzsche, is trying to escape from Hegel. [...] But to make a real escape from Hegel presupposes an exact appreciation of what it costs to detach ourselves from him. It presupposes a knowledge of how close Hegel has come to us, perhaps insidiously. It presupposes a knowledge of what is still Hegelian in that which allows us to think against Hegel; and an ability to gauge how much our resources against him are perhaps still a ruse which he is using against us, and at the end of which he is waiting for us, immobile and elsewhere."[51]

Paradoxically, both systems point to something beyond themselves. In Hegel the finiteness of all positions that are gone through indicates an "absolute spirit" which is realised in the finiteness of nature and history and ultimately returns to itself.[52] This explains the words at the beginning of the *Science of Logic*: this work represents God "as he is in his eternal essence before the creation of nature and a finite mind."[53] Mighty words indeed. Infinity does not oppose finiteness, but contains it as the "wealth of the particular."[54] Gödel and Turing construct a universal determinist system in order to demonstrate with the metaphor of paradox that freedom is conceivable within it. Already in 1928, Turing's teacher Godfrey Harold Hardy emphatically contradicted Hilbert's question concerning decidability:

---

49. Cf. titles of esoteric literature, such as Chiara Lubich, *Here and Now. Meditations on Living in the Present* (Hide Park, NY, 2000); or Ram Dass, *Be Here Now* (Three Rivers, MI, 1971).

50. Hegel, *Phenomenology*, p. 60f.

51. Michel Foucault, The order of discourse [1970], in: R. Young, ed., *Untying the Text. A Poststructuralist Reader* (London, 1981), pp. 48–78, here p. 74.

52. Hegel, *Phenomenology*, p. 493.

53. Hegel, *Logic*, p. 50.

54. Ibid., p. 58.

"There is of course no such theorem, and this is very fortunate, since if there were we should have a mechanical set of rules for the solution of all mathematical problems, and our activities as mathematicians would come to an end. […] It is only the very unsophisticated outsider who imagines that mathematicians make discoveries by turning the handle of some miraculous machine."[55]

However, Turing alone went on to deliver a strict disproof of this theorem by figuring out how to construct this "miraculous machine" with mathematical rigour.

---

55. Hodges, *Enigma*, p. 93f.

Traces of the Mouth
Andrei Andreyevich Markov's Mathematisation of Writing

A Short History of the Mathematisation of Writing

"There are very few things, which we know, which are not capable of being reduced to a mathematical reasoning; and when they cannot, it's a sign our knowledge of them is small and confused."

If this statement by the Scottish mathematician and physician John Arbuthnot, written in 1692, is correct, then our knowledge of the human use of language remains small and confused.[56] Despite the fact that numerous different cultures use the same symbols for letters and numbers, there is a deep divide between the two domains. Numerals can be expressed in words but not vice versa. For mathematics, language is a system of a non-describable order, although through probability calculations it has the ability to discover regularities even in highly chaotic data.[57] The belief in the universality and power of probability theory comes out very strongly in the following statement by Francis Galton in 1889:

"I know of scarcely anything so apt to impress the imagination as the wonderful form of cosmic order expressed by the 'Law of Frequency of Error.' The law would have been personified by the Greeks and deified, if they had known of it. It reigns with serenity and in complete self-effacement amidst the wildest confusion. The huger the mob, and the greater the apparent anarchy, the more perfect is its sway. It is the supreme law of Unreason. Whenever a large sample of chaotic elements are taken in hand and marshalled in the order of their magnitude, an unsuspected and most beautiful form of regularity proves to have been latent all along."[58]

---

56. Cited in Anders Hald, *A History of Probability and Statistics and Their Applications Before 1750* (New York, 1990), p. 183.

57. Its universality is due to its explanatory weakness. It does not attempt to postulate laws or forces, but only describes quantitative distributions.

58. Francis Galton, *Natural Inheritance* (London, 1889), p. 66. Thus the normal distribution can be seen as an empirical confirmation of Kant's concept of beauty.

However, it does not have the power to explain manifestations of subjectivity that are characterised by a radically different order. Since François Viète, letters had been used to denote numbers,[59] but numbers had never been used to denote letters. Before 1913, the theory of probability had little cognisance of the alphabet. In the early years, it was concerned with calculating the odds of winning at games of chance.[60] Another application was astronomy, as Anders Hald writes: "Observational and mathematical astronomy give the first examples of parametric model building and the fitting of models to data. In this sense, astronomers are the first mathematical statisticians."[61] John Graunt introduced the application of statistics to mass phenomena in a parallel development through his examination of death statistics in London between 1604 and 1662, with particular reference to the effects of the Black Death. This gave rise to later attempts at putting life insurance policies on a more reliable footing by calculating the probable life expectancy for different age groups.[62] Through the work of Adolphe Quetelet, Francis Galton, and Wilhelm Lexis, the application of statistics was expanded to include a large proportion of the phenomena of human life; not, however, to the social phenomenon of language.[63]

Before Markov, the symbols of the alphabet were only considered in two ways. First, in the period when their main subject was games of chance, they were the material for combinatorial calculations. In 1718, de Moivre interrogated their natural order in "Problem 35" of his *Doctrine of Chances*:

"Any number of letters a, b, c, d, e, f, & c. all of them different, being taken promiscuously as it happens: to find the probability that some of them shall be found in their places according to the rank they obtain in the alphabet; and that others of them shall at the same time be displaced."[64]

This derived from the convention of assigning capital letters to the various possible results of a sample. When their probability was calculated *a priori*, that is,

59. See François Viète, Introduction to the analytical art [1591], in: Jacob Klein, *Greek Mathematical Thought and the Origin of Algebra* (Cambridge, 1968), pp. 315–353. The founder of modern algebra was also a cryptologist; see Simon Singh, *The Code Book* (New York et al., 1999), p. 45.

60. See Hald, *History of Probability*, pp. 33–80. The main works in this connection are Cardano's *Liber de Ludo Aleae* (1564), the correspondence between Pascal and Fermat in 1654, Huygen's *De Ratiociniis in Ludo Aleae* (1657), and Jakob Bernoulli's *Ars Conjectandi* (1713).

61. Hald, *History of Probability*, pp. 144–182, who discusses Tycho Brahe's *De Nova Stella* (1573), Johannes Kepler's *Astronomia Nova* (1609), and Isaac Newton's *De Analysi per Aequationes Numero Terminorum Infinitas* (1669).

62. See Hald, *History of Probability*, pp. 81–143. Most important here are Graunt's *Natural and Political Observations Made Upon the Bills of Mortality* (1662), Witt's *Waerdye van Lyf-renten naer proportie van Los-Renten* (1671), and Halley's *An Estimate of the Degrees of the Mortality of Mankind* (1694).

63. See Adolphe Quetelet, *Sur l'Homme et le Développement de ses Facultés, ou Essai de Physique Sociale* (Paris, 1835); Francis Galton, *Hereditary Genius: an Inquiry into Its Laws and Consequences* (London, 1869); Wilhelm Lexis, Über die Theorie der Stabilität statistischer Reihen [1879], in: W. Lexis, *Abhandlungen zur Theorie der Bevölkerungs- und Moralstatistik* (Jena, 1903), pp. 170–212.

64. Abraham de Moivre, *The Doctrine of Chances: or, a Method of Calculating the Probability of Events in Play* (London, 1718), p. 109.

leaving aside empirical material, words appeared merely as exceptional cases of randomness. Each letter represented one selection among 26 equally possible choices and it was immaterial in which order they appeared. For this reason, in 1770, d'Alembert raised significant doubts about probability calculation as a whole:

> "In order to expand my idea with yet more clarity and precision, I suppose that we find on a table some printed characters arranged in this way: C o n s t a n t i n o p o l i t a n e n s i b u s, *or* a a b c e i i i l n n n n n o o o p s s s t t u, *or* n b s a e p t o l n o i a u o s t n i s n i c t n, these three arrangements contain absolutely the same letters: in the first arrangement they form a known word; in the second they form no word at all, but the letters are disposed according to their alphabetical order, and the same letters are found as many times in sequence as they are found in turn in the twenty-five characters which form the word; finally, Constantinopolitanensibus in the third arrangement, the characters are pell-mell, without order, and at random. Now it is first certain that, *mathematically speaking*, these three arrangements are equally possible. It is no less certain that all sane men who would cast a glance on the table where these three arrangements are supposed to be found, will not doubt, [...] that the first is not the effect of chance."[65]

That stochastically a meaningful word was just as probable as a meaningless one, ran counter to common sense. Six years later, in 1776, Laplace provided an answer to this problem. In a direct reply to the "very fine objections" of "Mr d'Alembert" he wrote:

> "Suppose that on a table we find letter types arranged in the order INFINITÉSIMAL; the reason which leads us to believe that this arrangement is not the effect of chance can come only from this that, physically speaking, it is less possible than the others, because, if the word *infinitésimal* were not used in any language, this arrangement would be neither greater, nor less possible, and yet we would suspect then no particular cause. But, since this word is in use among us, it is incomparably more probable that some person has thus arranged these letters than that this arrangement is due to chance."[66]

Laplace's argument eloquently glossed over the fact that for statistics, linguistic combinations were actually little more than very improbable combinations, and a person's choice of one word rather than another was merely a statistical phenomenon or an inexplicable fact. This branch of mathematical science could only state that a word is "in use among us"; it could not predict or understand it. In d'Alembert's reservations, however, there is a sense of what probability calculation could not grasp, or only very imperfectly: the order of language, including the individual that uses it.

65. Jean le Rond d'Alembert, Doutes et questions sur le calcul des probabilités [1770], in: *Oeuvres Philosophiques, Historiques et Littéraires de d'Alembert* (Paris, 1805), vol. 4, pp. 289–315, p. 305f. My translation – D.L.

66. Pierre-Simon Laplace, Recherches sur l'intégration des équations différentielles aux différences finies et sur leur usage dans la théorie des hasards [1776], in: *Oeuvres Complètes* (Paris, 1878–1912), vol. 8, pp. 69–197, cited in A. Hald, *A History of Mathematical Statistics from 1750 to 1930* (New York, 1998), p. 66.

The second application of numerals to letters developed in secret writing, or cryptology. Around 850 AD, the Arab scholar Al-Kindi described such a method:

> "One way to solve an encrypted message, if we know its language, is to find a different plaintext of the same language long enough to fill one sheet or so, and then we count the occurrences of each letter. We call the most frequently occurring letter the 'first', the next most occurring letter the 'second', the following most occurring the 'third', and so on, until we account for all the different letters in the plaintext sample.
>
> Then we look at the cipher text we want to solve and we also classify its symbols. We find the most ocurring symbol and change it to the form of the 'first' letter of the plaintext sample, the next most common symbol is changed to the form of the 'second' letter, and so on, until we account for all symbols of the cryptogram we want to solve."[67]

Known today as frequency analysis, this procedure makes use of the insight (which can only be gained *a posteriori*) that in empirical documents single letters are not equally probable, but appear with differing frequencies. This made early substitution codes, like the Caesar cipher, easy to crack. In this mathematisation, writing is a sequence of characters, which – if one takes sufficiently long sequences – appear in characteristic frequencies and thus remain recognisable in spite of their substitution with other characters. However, this method also did not permit differentiation between words that mean something and nonsensical words that contain the same letters. Up to the twentieth century, techniques for detecting regularities in cryptograms continued to be refined and eventually also considered the frequency of letter combinations.[68] However, Markov was the first to develop a complete theory that takes into account the connections between letters.

Markov's Mathematical Crosswords

On 23 January 1913, the Russian mathematician Andrei Andreyevich Markov gave a lecture to the physical–mathematical faculty of the Imperial Academy of Sciences in St. Petersburg.[69] The occasion was to celebrate the bi-centenary of the book that had formulated the fundamental theorem of probability: Jakob Bernoulli's *Ars Conjectandi*, virtually completed by 1690, but only published posthumously in 1713 by his nephew Niklaus because of family quarrels. However, in 1913, Russia was celebrating first and foremost a different anniversary: 300 years of the ruling house of Romanov. Obviously, this gave the physical–mathematical faculty's event a political dimension. Usually, mathematicians wrote texts about numbers, in which the former merely served to

---

67. Cited in Singh, *Code Book*, p. 35.

68. Around 1890, Etienne Bazeries used the frequencies of syllables to decrypt documents about the military campaigns of Louis XIV encrypted with the so-called "Great Cypher"; cf. Singh, *Code Book*, p. 57f.

69. Andrei A. Markov, An example of statistical investigation of the text *Eugene Onegin* concerning the connection of samples in chains [1913]. *Science in Context* 19. 4 (2006): 591–600.

elucidate the latter. Markov, however, shifted the periphery of his field to centre stage and lectured on calculations that he had performed upon a literary text.[70] In the years leading up to this, Markov had made theoretical extensions to probability theory, which traditionally only applied to independent events, to the field of trials that are dependent on each other. It was within a literary text that he finally found the material that would allow him to empirically verify his assumptions. His object of study was the most important work by renowned Russian writer Alexander S. Pushkin, *Eugene Onegin*. Applying the rigorous tool of statistics to this famous literary work, Markov went one step further than Pushkin, whose realistic style was directed against Romanticism, and delivered an analysis where signifiers were taken as the basis of all meaning. Soon afterwards, Markov published a second analysis of a literary work in the Appendix to the third edition of his *Probability Theory*: *Childhood Years of Bagrov's Grandson* (1858) by the lesser-known Russian author Sergei Timofeevic Aksakov.[71]

Both texts are autobiographical memoirs – Pushkin's work deals with the recollections of an 18-year-old and Aksakov's with memories of early childhood. Both writers conceal this by giving their main protagonists fictitious names: Eugene Onegin and Bagrov's Grandson. Memory binds the present moment to the past; there are no independent samples in life. Aksakov distinguishes between "fragmentary memories" (single images) and "connected memories," which were also the titles of the first two chapters.[72] A further circumstance that the two writers have in common is that they both wrote their works while living in isolation. Because of his political poetry, Pushkin was banished from St. Petersburg in 1820 and exiled to Kishinev in contemporary Moldavia. He had begun work on *Eugene Onegin*, when he was sent to Odessa (in today's Ukraine), and then a short time later to his mother's country estate at Mikhaylovskoe in Northern Russia. Aksakov's isolation resulted from the fact that he had been blind for ten years, and he dictated the memoir to his daughter.[73]

The passage from Aksakov's book containing 100,000 letters, which Markov analysed, describes a journey in a carriage to Orenburg, where the mother, who

---

70. This links his study with one undertaken eighteen years later by Kurt Gödel, which triggered a fundamental crisis in mathematics (Kurt Gödel, Über formal unentscheidbare Sätze der *Principia Mathematica* und verwandter Systeme, I. *Monatshefte für Mathematik und Physik* 38 (1931): 173–198). Both studies applied numbers to the material basis of signs, but Markov analysed texts written by others and thus avoided the self-reference that led to irresolvable paradoxes in the latter. Moreover, Markov investigated sequences of letters of the alphabet and not mathematical symbols.

71. A.A. Markov, On a remarkable case of samples connected in a chain [1924]. *Science in Context* 19. 4 (2006): 601–604. Surprisingly, although these calculations were more extensive and he conducted them around this time, Markov did not mention them at all in his letter to Chuprov dated 29 March 1916 (cf. Kh.O. Ondar, ed., *The Correspondence Between A.A. Markov and A.A. Chuprov on the Theory of Probability and Mathematical Statistics* (New York et al., 1981), p. 109).

72. Sergei T. Aksakov, *Years of Childhood* [1858] (New York, 1960), pp. 3 and 12.

73. On Pushkin, see Rolf-Dietrich Keil, *Puschkin* (Frankfurt a.M., 1999), p. 101f.; on Aksakov, see S.T. Aksakov, *Bagrovs Kinderjahre* [1858], trans. Erich Müller-Kamp (Zürich, 1978), p. 506: "Aksakov went blind in his left eye and the right eye began to flicker. By 1846, his sight had deteriorated so badly that he was hardly able to write his name." (My translation – D.L.)

is gravely ill, wishes to consult a doctor. The author lauds this form of transport, because it evokes the chain of one's biography:

"Travel – what a wonderful thing it is! Its powers are […] snatching him out of the environment he is in […], first directing his attention to himself, then to the remembrance of the past, and finally, to dreams and hopes that lie ahead."[74]

Under the aegis of the here and now, travel enables the human subject to step outside the sequence of the syntagmatic present and into thoughts of one's own biography; in this way, the subject is liberated from time.

A recurrent theme is the Russian language – Bagrov, the son of the family and the narrator, is an avid reader. His emotions while fishing are described thus: "With what a keen eye and what curiosity I gazed at those objects, new to me, how quickly I grasped their significance, and how easily and soundly I learned all their names!"[75] However, Russia is a multinational state and language is by no means uniform. Already at the beginning of their journey, the family encounters Chuvashes, Tatars, and Bashkirs. Statistically, this is to be expected as these groups represent the largest ethnic minorities. The family's servant attempts to adapt to the situation verbally: "Thinking to make himself more understandable, Evseich now began to distort his Russian frightfully, mixing in Tatar words. He wanted them to tell him where we could find worms for fishing." He gets the worms as well as the following answer: "Ai-ai! very much good fish catch here!"[76] Language is not presented as a static system, but rather as a structure that permanently changes, especially through contact with other languages and people's inability to find the right word. In *Eugene Onegin*, too, there are instances of language reshaping. In verse 35 of the first chapter, early one morning while the hero is still asleep, of course, at the German baker's the "васисдас" (transcribed in the Latin alphabet as "Wasisdas" – Whatsit) opens. The editor of the German translation of *Eugene Onegin* comments on this word, obviously derived from German, thus: "This word form was introduced in Russia during the foreign invasion 1812 and was used to denote the little flapping or sliding windows of the German bakeries in Petersburg."[77] In French, the loan-translation "vasistas" is documented since 1798.[78]

In these texts, Markov investigated the frequency of vowels and consonants and the possible connections between them. Since his research on Pushkin is more detailed, I shall focus on this paper and only mention his analysis of Aksakov's text where the techniques or results differ. Markov divided the first 20,000 letters of Pushkin's novel into 200 groups of one hundred letters and wrote each group in a square table with ten rows and ten columns. Not included were the hard signs (ъ) and soft signs (ь), which are not pronounced independently but modify the pronunciation of the preceding letter, punctuation, and spaces.

74. Aksakov, *Years of Childhood*, p. 53.

75. Aksakov, *Years of Childhood*, p. 24.

76. Aksakov, *Years of Childhood*, p. 30f.

77. Alexander Puschkin, *Gesammelte Werke*, ed. Harald Raab (Frankfurt a. M., 1973), vol. 3, pp. 24 and 436. My translation – D.L.

78. See A. Pushkin, *Eugene Onegin* [1833], trans. Vladimir Nabokov (Princeton, 1981), vol. 2, p. 145.

He counted the vowels in each column, combined two at a time (the first and the sixth, second and seventh, third and eighth, fourth and ninth, fifth and tenth), and wrote down the five sums underneath each other in a vertical column. Thus, each sum represented the number of vowels among 20 letters that were separated by four letters in the text. Combined, they showed the number of vowels in a sequence of 100 letters of the novel. Markov was able to represent the entire text with 40 tables, each with five of these columns representing 500 letters from the novel, and all of this filled a whole page. Here, he first added the numbers vertically and entered the total at the bottom in boldface. The square matrix of the table also permitted addition of the numbers horizontally and thus a calculation of how many vowels there were per 100 letters that were separated by four letters in the text. He entered these totals, also in bold, in the final column. Finally, Markov added the vertical and horizontal sums together and, to save space, entered the total minus 200 in the bottom right-hand corner of each table. As it was the same 500 letters that were being counted using different methods, the addition of the last lines and last columns gave the same result.

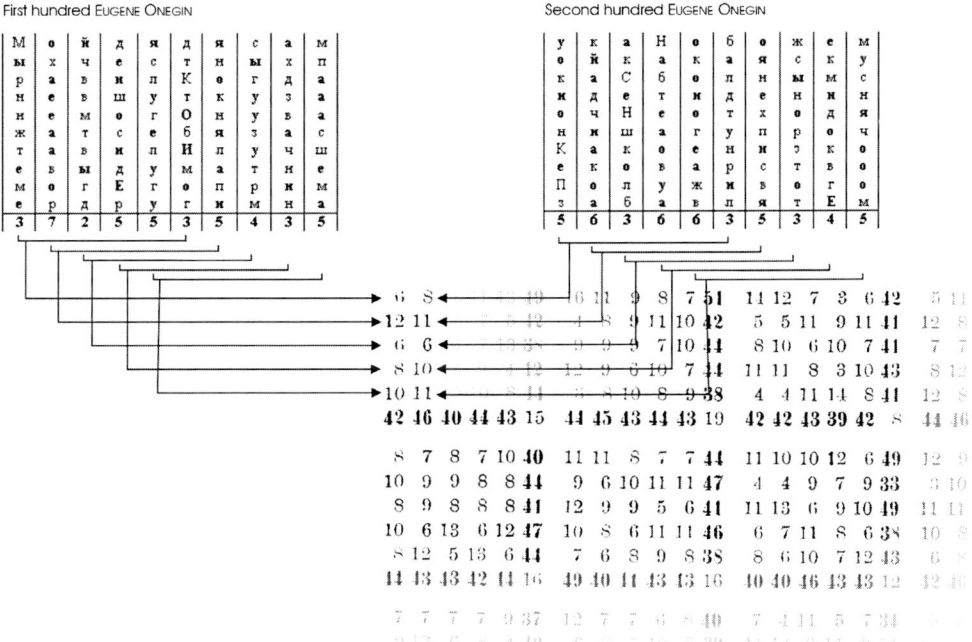

Fig. 1: Markov's counting method.

The text was broken up according to a specific scheme – at every tenth letter. This imitated the text's form, that of a novel in verse. The existing caesuras in the language were replaced by the regular and artificial ones of mathematical counting. Originally, the Latin word *versus* meant the turn of the plough at the end of a furrow and derives from the verb *vertere*, to turn. In contrast to the style of "prose,"[79] where the only direction is forward, the flow of letters in a poem – particularly in rhyming verse – returns constantly to the beginning of the line. This facilitates a second reading of the text in addition to the syntagmatic one: one that breaks its linearity. The written form of language inherits linearity from the spoken word and, at the same time, integrates it, since on book pages, stone tablets, or papyrus rolls, the acoustic thread along which perception makes its way is transformed into two dimensions. In a similar way to how the novel's traveller is liberated from being a slave to time on the journey and can think in free associations, the written form gives the reader the freedom to choose his or her own way through the rectangular field of letters, to go backward or forward, skim through, or read every which way. Unlike speaking, writing is not a linear medium.[80]

The constraints exerted by orality are two-fold and concern the organs of the mouth and the ear. In the case of that which is mistakenly taken for the transmitter, this means that without exception, only one word at a time – and thus one word after another – can be spoken. Jacques Derrida termed this restriction of speech "angustia."[81] A prerequisite for progress in a straight line is the Hegelian fading away ("Verklingen") of the voice, in other words, the fact that the medium – air – cannot store information.[82] Only an echo realises memory, in a way similar to how the engineers of the first computers created it: as a self-propagating impulse that is continually refracted and mirrored between two points in a liquid carrier medium, and it is through this continual echo that it survives.[83] The nightmare of a continually echoing space, which, since the 1960s, is possible to create using taped sound loops and other special effects devices, allows the opposite of a soothing, idealistic fade – the expression of just one single sound that will continue to sound for all eternity and is thus stored.[84] Because of its constraints,

---

79. Latin *prorsus* – forward, straight ahead, to sum up, utterly, wholly.

80. For a different point of view, see Jay David Bolter, *Writing Space* (London, 1991), p. 108: "But in most books, as in the papyrus roll, one path dominates all others – the one defined by reading line by line, from first page to last. The paged book has a canonical order."

81. Jacques Derrida, *Writing and Difference* [1967], trans. Alan Brass (London, 1993), p. 9: "*angustia:* the necessarily restricted passageway of speech against which all possible meanings push each other, preventing each other's emergence."

82. Cf. Georg Wilhelm Friedrich Hegel, *Encyclopedia of the Philosophical Sciences in Outline and Critical Writings* [1830], ed. Ernst Behler (New York, 1990), p. 233: "But since it [the intuition] exists only as suspended, and the intelligence is its negativity, the true form of the intuition as a sign is its existence in time, – but this existence vanishes in the moment of being, and its tone is the fullfilled manifestation of its self-proclaimed interiority."

83. See Alan M. Turing, Lecture to the London Mathematial Society on 20 February 1947, in: D.C. Ince, ed., *Collected Works of A.M. Turing* (Amsterdam et al., 1992), vol. 1, pp. 106–124, p. 109: "The idea of using acoustic delay lines as memory units [...] The idea is to store the information in the form of compression waves travelling along a column of mercury. [...] [I]t is quite feasible to put as many as 1000 pulses into a single 5' tube. [...] A train of pulses or the information which they represent may be regarded as stored in the mercury whilst it is travelling through it. If the information is not required when the train emerges it can be fed back into the column again and again until such time as it *is* required."

84. The first machine of this kind was put on the market at the end of 1958 by Hans Bauer. It was named Echolette NG 51/S.

the field of what is audible has either no or only one storage space. During actual experience, time is zero-dimensional, because its expansion is only present in its passing. That which is voiced by the mouth disappears immediately in the general medium or silences everything – *out of memory*.

This restriction applies equally to the ear. A listener cannot re-wind or fast-forward the medium of transmission. It has no storage capability and there is no way for him to access it later. Thus, he is obliged to remain a slave to the constant now of speech, which runs by him. Sorry, what did you say?

The "return" breaks the linearity of Chronos on and in space. Language has always been set down in the form of cut strips. The make-up of text in a second dimension probably took place ca. 1500 BC and this is already reflected upon in the Bible. There are numerous instances of acrostics, where the initial letters of the lines form the complete alphabet and were an aid, for example, in the Psalms, in committing texts to memory.[85] Besides mnemotechniques, others that use proper names served as dedications to other persons, self-praise by the author, or for concealing messages in general. With the advent of computers and word processing programmes, the function of the "return" key changed from its original one of line advance and return to the line's beginning into the final confirmation of a command.[86]

The temporality of speech set down on a book page was not the only precondition of Markov's operation. Columns are formed only when the characters are distributed discretely and regularly spaced on the page's area. This is a quality that oral expression lacks and, to a certain extent, also hand writing. The organic flow protects the line from being read vertically. The first typewriter with Cyrillic letters, the Model 8, began being exported by the German Adler Company in 1903.[87] Like cryptography, the mathematician's studies are a writing game with discrete characters.[88]

The expansion of language into the second dimension, which is effected by the transcription of the voice into writing, Markov implemented as a paper-automaton and employed it to calculate "other combinations"[89] within

---

85. For example, no. 119 of the *Tehilim* (Psalms), in which eight verses each begin with the same letter and 176 lines run through all 22 letters of the Hebrew alphabet.

86. Cf. Chapter 7, which details a very early instance of "enter." In Joseph Weizenbaum's ELIZA of 1966, one of the first dialogue programmes that linked questions from a human interrogator to answers from a computer with different probabilities, it had the following function: "The user's statement is terminated by a double carriage return, which serves to turn control over to ELIZA." (Joseph Weizenbaum, ELIZA – a computer program for the study of natural language communication between man and machine. *Communications of the ACM* 9 (1966): 36–45, p. 36) In the pseudo-oral communication via terminals it set a caesura, but in this case a caesura between two partners in a dialogue. Thus, on the computer keyboard there is not only the "return" or "enter" button, but also a "home" button, which, however, does not make the line advance forward.

87. See Friedrich A. Kittler, *Discourse Networks 1800/1900* [1985] (Stanford, 1990), p. 193: "Spatially designated and discrete signs – that, rather than increase in speed, was the real innovation of the typewriter."

88. The only known example of oral cryptography was used by the Navajo Division of the US Army during the Second World War. See Singh, *Code Book*, p. 196f.: "After three weeks of intense cryptanalysis, the Naval codebreakers were still baffled by the messages. They called the Navajo language a 'weird succession of guttural, nasal, tongue-twisting sounds […] we couldn't even transcribe it, much less crack it.'"

89. Markov, Example of statistical investigation, p. 592.

the text material with the aim of bringing out the properties of the natural syntagmatic sequence.

Markov began by focusing on the results of the consecutive calculation. A table shows how often the numbers from 37 to 49 occured in the final rows of the small tables, which contained the number of vowels in the sequences of 100 letters. The modal, in statistics, the most frequent value, was 43. From this, Markov calculated the arithmetic mean. He multiplied the differences to the modal with the frequency of their occurrence, added the results, and divided this by the total of tests. The mean deviation from the most frequent number thus obtained was added to this and gave the mean sought, 43.19, as the average number of vowels in 100 letters. The unusual method of deriving this from the modal had the advantage – for someone doing calculations by hand – of working with much smaller quantities than is customary today, by multiplying the values themselves. After dividing by 100, p – the mean probability of any single letter being a vowel – was approximately 0.432.[90]

However, the actual number of vowels in the various groups of 100 letters differed. In mathematics, the value indicating how far single numbers deviate from their mean is called "dispersion." To determine this dispersion, Markov used a technique, which was first published in 1805 by Legendre in *Nouvelles Méthodes pour la Détermination des Orbites des Comètes* but is, however, linked with Gauss' name, who elucidated it in 1809 in his *Theoria Motus Corporum Coelestium*: the method of least squares.[91] The deviation of each value from the arithmetic mean of 43.19 was squared, and thus positive; next, the average of these differences was calculated by multiplying them by their frequency of occurrence, adding them, and finally, dividing them by their total number. The sum of the squares of deviations was 1022.8; division by 200 gave 5.114. By drawing the root from the result, one usually arrives at the so-called "standard deviation" as a relative measure of the dispersion. The mathematician, however, simply worked with the square of this value, the so-called "variance." The number of vowels per 100 letters thus differed on average approximately $\sqrt{5.114} = 2.26$ from its mean.[92]

---

90. In Aksakov's text, however, p was 0.44898 and was thus slightly higher. This is rather surprising, for one would expect there to be more vowels in poetry than prose. Cf. Markov, Remarkable case, p. 602.

91. Adrien-Marie Legendre, *Nouvelles Méthodes pour la Détermination des Orbites des Comètes* (Paris, 1805); Carl Friedrich Gauss, *Theoria Motus Corporum Coelestium in Sectionibus Conicis Solem Ambientum* (Hamburg, 1809).

92. Also when determining the standard error of the mean, Markov calculated its square, by dividing the variance 5.114 by the number of samples, 200. Normally, the mean dispersion was divided by the square root of this value. In 1812, Laplace discovered that the dispersion of means from samples, which were all taken from the same group, turned into a normal curve with growing number. Through the so-called Central Limit Theorem, in dependence on the quantity of samples, it was possible to calculate the expected deviation from the real average. Because the errors were normally distributed, it was possible to define confidence intervals around the found value, within which the actual average of the sum total would certainly lie. 96% of the samples were approximately within the range of the two-fold standard error. Cf. P.-S. Laplace, *Théorie Analytique des Probabilités* (Paris, 1812). With 96% certainty, it could be said that the real average of vowel frequency in Pushkin's novel lied between 42.87 and 43.5. The standard error was around 0.16.

Language as a Random Process

The method of least squares, traditionally and until the work of Markov, could only be legitimately applied to quantities that were independent of each other. It held true in this case, because the number of vowels in the first 100 letters had virtually no influence on how many there were in the second 100. There was only dependence between the last letters of the first batch and the first letters of the second since they were considered in their linear sequence. This was why the dispersion followed the normal distribution – the first surprising discovery presented in Markov's lecture. In 1733, de Moivre developed the normal distribution in connection with problems relating to games of chance. It approximated the expected binomial distribution for independent random elementary errors, which was complicated to calculate for large numbers of samples.[93] As a result of its properties being known, it was possible to determine the probability that values were located in certain intervals. For example, half of them usually fell in the range of the 0.6745-fold of the standard deviation to the left and the right of the arithmetic mean. This discovery was used for the first time in 1815 by Friedrich Wilhelm Bessel, director of the Royal Observatory in Königsberg (now Kaliningrad), in his studies on the position of the pole-star to estimate errors in astronomical measurements and was termed "probable error."[94] Thus, it could be assumed that 100 of Markov's 200 values were $2.26 \times 0.67 = 1.5$ away from the average. And actually there were 103 values between $43.2 - 1.5 = 41.7$ and $43.2 + 1.5 = 44.7$, which was a good fit with the theory.

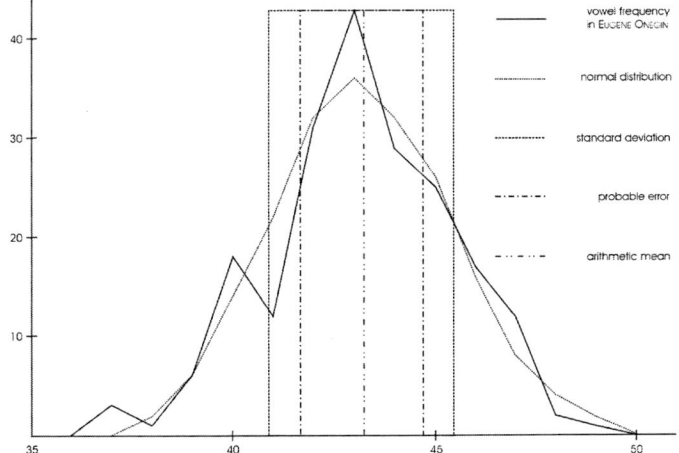

Fig. 2: The values of the horizontal calculation compared to the normal curve.

93. A. de Moivre, Approximatio ad summam terminorum binomii (a + b)ⁿ in seriem expansi [1733], in: A. de Moivre, *The Doctrine of Chances*, 2nd ed. (London, 1738), pp. 243–259.

94. Friedrich Wilhelm Bessel, Ueber den Ort des Polarsterns, in: Astronomische Jahrbücher für das Jahr 1818 (Berlin, 1815), pp. 233–241.

The method of least squares can help to correlate theoretical models with empirical data, for example, geometrical measurements in the case of geodesy, and in this way it removes the instruments along with their inevitable inaccuracies from the data.

> "However carefully one takes observations of the magnitudes of objects in nature, the results are always subject to larger or smaller errors. [...] Such errors come from the imperfections of our senses and random external causes, as when shimmering air disturbs our fine vision. Many defects in instruments, even the best, fall in this category; e.g., a roughness in the inner part of a level, a lack of absolute rigidity, etc."[95]

The law of error allowed empiricism to be corrected by theory, in that it converted numerically measured values into possible ranges of results and thus weakened them. Observable quantities were typically continuous and thus could be made more precise and approximated ad infinitum. By contrast, Markov's material was precisely defined on the level that he was dealing with. Obviously, printed letters exhibit certain irregularities, but this did not concern this enquiry. Markov did not make measurements; he counted various discrete entities. There might be some mistakes in his tallies, but they were not thematic. Any deviation in the number of vowels from the average was simply understood as an error. Whereas in other cases it was a question of using the Gaussian method to make the real fit closer to the symbolic, here the symbolic was taken as the real and its internal irregularity re-interpreted as an error of measurement.

That the distribution of the number of vowels in the groups of 100 letters corresponded to a random dispersion is something that demands an explanation. For language is certainly not determined by the contingency of an individual wielding tools, but is the intentional expression of a human subject. The subject can act within the oral space freely and without resistance so that he even masters it without a feedback loop. The accident of real collisions or mishaps, as evidenced by measurement errors, is precisely that which the oral space excludes in an incomparable manner.[96] Why is language statistically subject to the same contingency as the meeting of physical objects with an observer or balls falling in the Japanese game of Pachinko?[97] This might be plausible for deviations in pronunciation, but not for counts of discrete letters of the alphabet. Markov simply explained his astounding findings by way of the independence of the samples. The number of vowels in the first batch of 100 letters does not determine how

---

95. C.F. Gauss, *Theory of the Combination of Observations Least Subject to Errors* [1821–1828] (Philadelphia, 1995), p. 3.

96. See Detlef B. Linke, Die Vorläufigkeit der Neurotheologie. *Lettre International* 61 (2003): 106–107, p. 106: "The oral space is therefore a space that [...] exhibits predictable parameters which are relatively stable, that is, in the case of its own movements, so that sensory feed-back is not necessary. [...] Instead of feed-back, there can be combinations of sounds and their ordering under new laws, those of grammar. According to this model, the oral space provides the security that the outside world cannot guarantee." (My translation – D.L.)

97. This game, popular all over Japan since the 1920s, bears a remarkable resemblance to Francis Galton's random generator "Quincunx," which he developed in 1877 to illustrate the normal distribution. Cf. F. Galton, Typical laws of heredity. *Nature* 15 (1877): 492–495, 512–514, 532–533.

many there will be in the second. This explanation, however, is not sufficient to explain the observed normal distribution. That the frequencies are dispersed in this manner forces one to advance the radical thesis that the source of language is a random process. This is not to be expected. In their critique of the application of the normal curve to social and psychological phenomena, Fashing and Goertzel, who think that it serves as the foundation of a theory of social inequality, write:

"[T]he normal bell curve is 'normal' only if we are dealing with random errors. Social life, however, is not a lottery, and there is no reason to expect sociological variables to be normally distributed."[98]

At the level of the speaking or writing human individual, the random theory does appear completely absurd, because people implement their uninhibited intentions in such actions and say what they want to say. However, if one looks at the social aspect – that if one wants to be understood, one has to use the language everyone else is using – then there is an explanation. Although Pushkin was free to choose what he expressed and which chains of letters he strung together, he utilised a repertoire of well-formed words that was already established at the time he was writing. Moreover, the manner in which words were put together was to a certain extent governed by syntactic and grammatical rules. Pushkin, or anyone, expressed personal thoughts in a general medium. In a text there are two forming processes at work, which differ considerably in their temporality, spatiality, and origination: the collective process of language formation over time and, conditioned by this, the individual process to articulate specific contents. At first, one is unable to decide where the randomness might come in.

The following experiment provides clarification in this matter. From an alphabetical list of Russian words, as used by hackers and system administrators to test the safety of passwords, random entries are chosen until the total number of characters reaches 20,000.[99] The result is a completely nonsensical string of valid expressions in Russian, devoid of all syntax or grammar. The act of an author, which possibly produces randomness through selecting certain words to formulate a specific content, is omitted. The sample thus produced is then examined using Markov's method. The following table (Figure 3) gives the results.[100]

---

98. Joseph Fashing and Ted Goertzel, The myth of the normal curve. *Humanity and Society* 5 (1981): 14–31, p. 27; see also p. 16: "The bell curve came to be generally accepted, as M. Lippmann remarked to Poincaré […], because […] the experimenters fancy that it is a theorem in mathematics and the mathematicians that it is an experimental fact."

99. The list used is the file russian_words.koi8.gz available at http://www.funet.fi/pub/unix/security/passwd/crack/dictionaries/russian/. The source code of the programme can be accessed at http://alpha60.de/research/.

100. The values deviate slightly from those in Markov's study, because a modern version of Pushkin's text was used for this comparison.

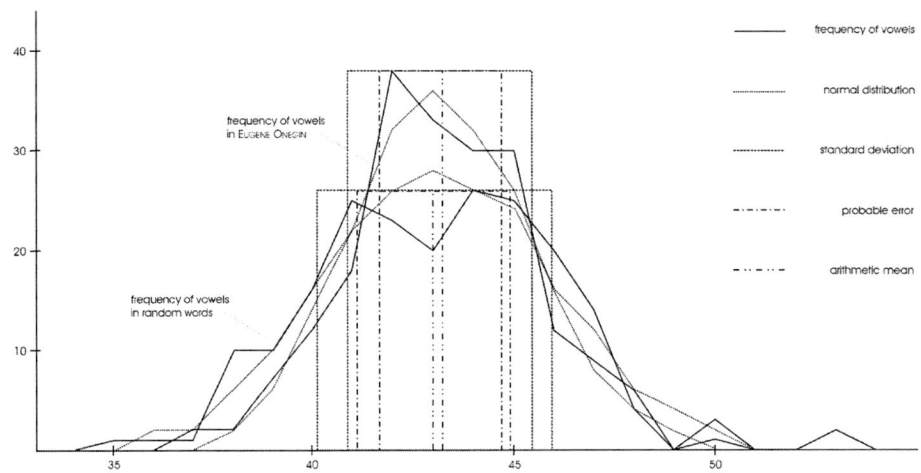

	arithm. mean	sum of the squares of the deviations	standard error	range of error	number of values in this range
EO	43.19	1,045.3	1.5	41.6 – 44.7	101
RW	43.06	1,702.3	1.9	41.1 – 45.0	94

Fig. 3: Comparison of the frequency of vowels in Eugene Onegin (EO) and random words (RW).

The results of the calculations performed on the random words hardly differ in their proximity to the normal curve from those of Markov on *Eugene Onegin*. This applies equally to all of Markov's other calculations. Thus, they do not concern a particular author's style and manner of arranging words but language in general. The conclusion, therefore, from the dispersion of values is that the process of formation, which generates the ensemble of all allowable chains of letters, is a random process.

Since Markov made no comment whatsoever on this randomness, we must search for an explanation in the linguistic theory of his time. From 1907 to 1911, the Swiss linguist Ferdinand de Saussure held a series of lectures at the University of Geneva, which were later published posthumously by his students as *Course in General Linguistics* and became one of the most influential language theories of the twentieth century. As far as we know, there was never any direct contact between the Russian mathematician and the Swiss linguist, nor did they know each other's work.[101] Therefore, the close correspondences described

---

101. According to Paul Bouissac, the *Course* was only introduced into Moscow around 1917, when the Russian linguist Serge Karcevski returned from Geneva where he was in contact with Charles Bally, one of the editors of Saussure's lectures. He emigrated to Switzerland in 1907; cf. Paul Bouissac, Perspectives on Saussure (2003; online: http://www.semioticon.com/people/articles/saussurecompanion.rtf.htm).

below should be considered as a kind of "parallel invention" that sheds considerable light on the meaning of Markov's calculations.

Saussure compared diachronic changes of language to chess and established the following agreements:

> "(a) One piece only is moved at a time. Similarly, linguistic changes affect isolated elements only.
>
> (b) In spite of that, the move has a repercussion upon the whole system. It is impossible for the player to foresee exactly where its consequences will end. [...]
>
> (c) [...] In a game of chess, any given state on the board is totally independent of any previous state of the board. It does not matter at all whether the state in question has been reached by one sequence of moves or another sequence."

In the following passage, he declared the process of language formation to be random:

> "There is only one respect in which the comparison is defective. In chess the player *intends* to make his moves and to have some effect upon the system. In a language, on the contrary, there is no premeditation. Its pieces are moved, or rather modified, spontaneously and fortuitously. [...] If the game of chess were to be like the operations of a language in every respect, we would have to imagine a player who was *either unaware of what he was doing or unintelligent.*"[102]

The origin of these spontaneous changes lies in changes to spoken language initiated by a few people:

> "*[E]verything which is diachronic in language is only so through speech.* Speech contains the seed of every change, each one being pioneered in the first instance by a certain number of individuals before entering into general usage. [...] But not all innovations in speech meet with the same success."[103]

The question why these people begin to speak differently "is one of the most difficult tasks in linguistics" and cannot be explained completely.[104] "[B]lind forces of change" have an effect on the organisation of the system of signs.[105] As the main source of mutations, Saussure identified changes to the sound of words, to which there are no limits.

---

102. Ferdinand de Saussure, *Course in General Linguistics* [1916], ed. C. Bally and A. Sechehaye (London, 1983), p. 88f. My emphasis – D.L.

103. Saussure, *Course*, p. 96f.

104. Saussure, *Course*, p. 146.

105. Saussure, *Course*, p. 89. This formulation demonstrates that Saussure was close to the idea of the autonomy of language, as advocated later by Lacan and Burroughs. "These linguistic features were spread by contact" (Saussure, *Course*, p. 205) is echoed in Burrough's "blind forces" of language which he identifies as a viral process.

"If one attempts to evaluate the effect of these changes, one soon sees that it is unrestricted and incalculable. It is impossible, in other words, to foresee where they will end."[106]

For Saussure, the "most important factor of change" was "imagination over a gap in memory."[107] When speakers do not know how to answer the above question about the "wasisdas," they transform the language and make modifications. These are spontaneous in the sense that they do not follow a particular goal. They do not know whether their modifications will be taken over by others or what further effects they may elicit. Such actions are not purposive; nor can their causality be predicted. It is, therefore, plausible that in total this results in a random distribution. A further point against the above-mentioned caveat by Fashing and Goertzel – not to apply the normal curve to social phenomena – is that even collective processes, where the actors possess greater clarity regarding the goals and effects of their actions, result in Gaussian dispersions, for example, events on the stock exchange.[108]

The randomness of this genesis leaves its marks on the chains of letters, which can be discerned in Markov's calculations but which he did not comment on. The findings of the mathematician would have pleased the linguist Saussure, who described language as a "mechanism, which involves interrelations of successive terms," "like the functioning of a machine in which the components all act upon one another even though they are arranged in one dimension only."[109]

Constraints on Randomness

Markov then turned his attention from the groups that were independent of each other to the dependence of the single letters in these groups. He analysed the dispersion on this level to see whether the connectedness had an influence on it. He calculated the deviation of each letter from the mean probability of being a vowel (0.432) and then divided the variance of the sample (5.114) by the number of letters it contained (100). The result was 0.05114. For independent events, however, one would theoretically expect a considerably higher dispersion. Random samples, which are independent of each other and only have two possible outcomes, approach the binomial distribution. The expected variance is calculated from the multiplication of the probabilities and, in this particular case, was $0.432 \times 0.568 = 0.245376$. When the theoretical result was compared with the empirical result derived from the text, the coefficient of dispersion, which relates the two values, was 0.208. Thus, the variance of every single letter was five times smaller than one would expect.

106. Saussure, *Course*, p. 150.

107. F. de Saussure, *Linguistik und Semiologie* (Frankfurt a.M., 1997), p. 431. My translation – D.L.

108. See Louis Bachelier, Théorie de la spéculation. *Annales Scientifiques de l'Ecole Normale Supérieure* 17 (1900): 21–86. He explained the fluctuation of the value of stocks with the concept of Brownian motion.

109. Saussure, *Course*, p. 127.

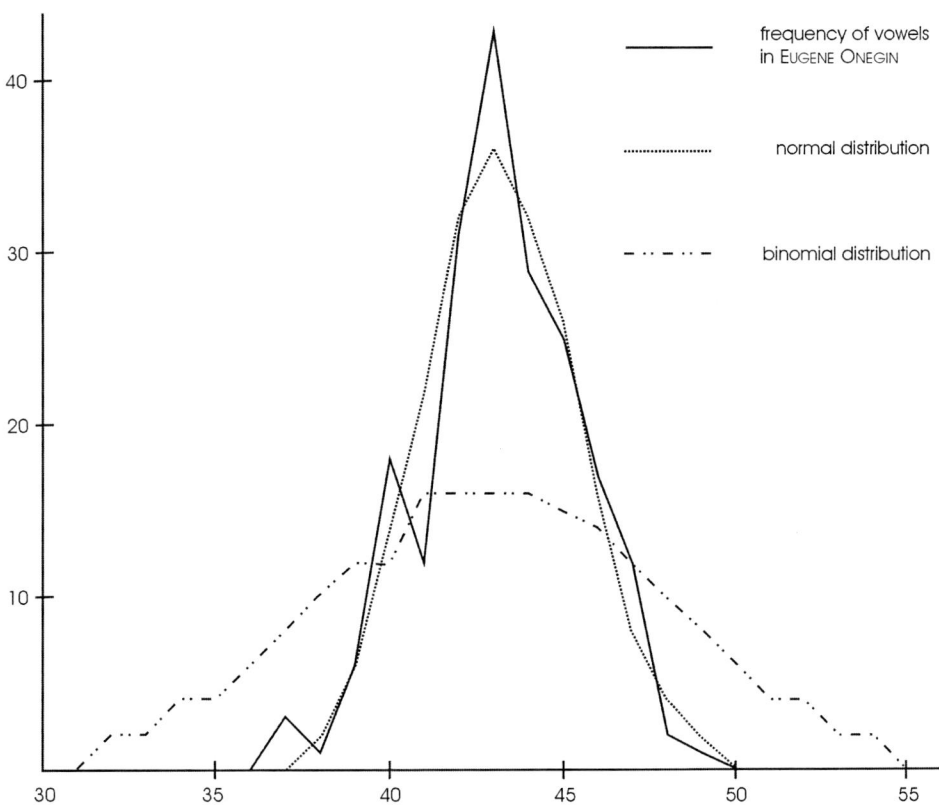

frequency of vowels
in EUGENE ONEGIN

normal distribution

binomial distribution

*Fig. 4: The frequency of vowels in Eugene Onegin (EO)
compared to the binomial distribution.*

This method of analysing the stability of statistical series was developed in 1879 by Wilhelm Lexis, a German mathematician, social scientist, and economist. Dissatisfied with the contemporary common practice in statistics, whereby material was assumed uncritically to be normally distributed, Lexis was the first to compare the empirically found standard deviation with the theoretically predicted deviation and to form their coefficient Q. When the two corresponded, that is, when Q = 1, Lexis spoke of a "normally random distribution": "It evolves only because a constant base probability becomes apparent with only the incertitude in the observed numbers which is admissible by analogy with a correct game of chance."[110] He judged such a series to be maximally stable in respect to the underlying probability. However, if the observed dispersion was higher than the theoretical dispersion, Q was greater than 1. Lexis termed this "supernormal" and explained it with "the fact that the 'normal random' fluctuations of the base probabilities combined with the physical ones," that is, changes of the probabilities themselves. For an "undernormal" dispersion, where Q was smaller

---

110. Lexis, Theorie der Stabilität, p. 183. My translation – D.L.

than 1, as with Markov, Lexis gave the verdict that "it would indicate that the mass phenomenon under consideration is internally connected or subject to certain regulating interventions or norms. It would more or less belong to the field of systematic order or commanding laws," that is, not that of probability calculation.[111]

This was precisely the case regarding single letters, which we are looking at here. In the first counting method, consecutive series of 100 letters formed groups, which were dependent on one another in the sense that a certain number of preceding consonants forced the next letter to be a vowel and vice versa. This led to the circumstance that their total number fluctuated much less than lots drawn in a game of chance with constant probabilities. Unlike Lexis, who banned this area from the magisterium of stochastics, Markov attempted to grasp it mathematically.

To this end, he counted the frequencies of the sequences vowel–vowel and consonant–vowel. It turned out that the probability of a vowel varied according to which type of letter preceded it. When this was a consonant, it was 0.663; when a vowel, only 0.128. Since 1906, Markov had studied simple chains as groups of samples in which each member determined the next; however, entirely theoretically due to a lack of suitable empirical material.[112] Using an equation from his *Investigation of a Notable Case of Dependent Samples*, Markov calculated from the difference δ of the two values, which he called p_0 and p_1, a theoretical coefficient of dispersion of 0.3, which was very close to the empirical one of 0.208, and in this way confirmed for the first time his mathematical deductions experimentally.[113] Thus the empiric dispersion could be explained much better if one assumed that in Pushkin's text the sequences of letters were chains rather than independent entities.[114] In the "simple chain" that he postulated for the purpose of approximation, each element was dependent on the preceding one.[115]

---

111. Ibid. My translation – D.L.

112. The only example of connected samples in probability theory before Markov was the experiment in which balls were taken out of an urn and not put back. It had the sole function of illustrating *a priori* findings afterwards. The calculations were not tested empirically because no one saw any reason to doubt them.

113. A.A. Markov, Recherches sur un cas remarquable d'épreuves dépendantes. *Bulletin de l'Académie Impériale des Sciences de St.-Pétersbourg* 1. 16 (1907): 61–80. In Russian; slightly different version published in French in: *Acta Mathematica* 33 (1910): 87–104; the equation used is on p. 100 in the latter version.

114. In Aksakov's text, too, the dispersion coefficient of 0.25 was clearly "subnormal." Here, the theoretical value for the simple chain is 0.29.

115. Around twenty years later, Alan Turing proved that for the hand-written calculations, with which the Russian developed his findings in lengthy and wearisome counting work, the second dimension was not essential. Turing then proceeded to develop his idea of a universal machine, which to a large extent replaced the "computers" of the time. In this construction, too, each sign was dependent on the preceding one; mediated, however, by the state of the machine and the programme that linked them. Additionally, the connection was no longer probabilistic, but completely defined. The decisive difference, however, is that Turing's machine – in contrast to Markov's, which always moved towards the right – returned to what it had written and interpreted that as a command. It re-wrote it and thus changed its own programme. The algorithm of the Russian mathematician wrote readable signs, Turing's wrote executable signs. For this reason, the former did not possess a halting condition. (The halting condition denoted the coincidence of state and read sign, when the universal machine turned off and delivered a result.) Cf. A.M. Turing, On computable numbers, with an application to the Entscheidungsproblem. *Proceedings of the London Mathematical Society (Ser. 2)* 42 (1937): 230–265.

However, the value could be even better approximated following the hypothesis that each letter was dependent on the pair of letters preceding it. Markov investigated this in his paper of 1911, *About a Case of Samples Connected in Multiple Chain*. He counted the frequency, $p_{1,1}$, of sequences of three vowels and of three consonants, $q_{0,0}$, in Pushkin's text. After inserting these values in an equation from the above paper, he found the theoretical coefficient of dispersion to be 0.195, which was even closer to the empirical coefficient of dispersion.[116] The first arrangement, therefore, gave 200 nearly independent tests each with 100 enchained letters.[117]

The dispersion was normally distributed; however, at the level of individual letters it was five times smaller than would be expected for a random process with this mean. To employ the main metaphor of stochastics, this resembled "an irregular game where the results of the single series of experiment are pushed closer to the mean value as would be expected in a game of chance with constant probabilities by voluntary influence. Typical quantities of this sort do not belong to the quantities of probability but are only of the same form."[118] Language appeared as a game with a marked deck of cards. The random generator was formed atypically in such a way that it produced a distribution that was too constrained. Similar methods can be utilised to detect manipulated games of chance without having to take the mechanism, for example, roulette, into account. The case found here was incomprehensible to probability theory, because the oscillation of the probabilities themselves cancelled out the Gaussian noise. Lexis underlined this by demonstrating that in his equation for calculating the total dispersion, $R = \sqrt{r^2 + p^2}$, when this was smaller than the random dispersion, r, the factor p, which gave the physical deviation, became imaginary.[119]

Reasons for the mathematical limitation of variance can also be found in Saussure's lectures. He not only described the ceaseless proliferation of signifiers through the modification of sounds, but also phenomena that, in turn, constrained this:

---

116. A.A. Markov, Sur un cas d'épreuves liées en chaîne multiple. *Bulletin de l'Académie Impériale des Sciences de St.-Pétersbourg* 5. 2 (1911): 171–186. In Russian; the equation used is on p. 179.

117. That the samples examined were independent was also shown by the fact that the dispersion varied only insignificantly when they were combined in twos, fours, or fives. If chaining would have existed, this procedure would have made it decrease because in the groups formed fewer letters would have been mutually dependent. Dependency would have wandered into the sample, so to speak, and would have affected in decreasing measure one letter's relationship to the next. For the calculation in pairs Markov added two sums next to each other in the last row and in this way arrived at the number of vowels in the series of 200 letters of the text. He found the sum of the squares of the deviations from the double mean 86.4 to be 827.6, which did not differ significantly from the number for the hundreds, 1022.8. Similarly, when this was calculated for 50 groups of 400 letters or 40 groups of 500 letters, the variance remained stable: the former was 975.2 and the latter 1004.

118. W. Lexis, *Zur Theorie der Massenerscheinungen in der menschlichen Gesellschaft* (Freiburg, 1877), p. 34. My translation – D.L.

119. See Lexis, Theorie der Stabilität, p. 177: "Whereas under the presuppositions made here the inequality R < r can never arise because in this case p would become imaginary which points at an impossibility." (My translation – D.L.) The equation Q = 1 was discussed repeatedly by Markov and Chuprov in their correspondence as of 23 November 1910 (Ondar, Correspondence, p. 38f.). In their view subnormal dispersion was possible also in independent samples.

"For the entire linguistic system is founded upon the irrational principle that the sign is arbitrary. Applied without restriction, this principle would lead to utter chaos. But the mind succeeds in introducing a principle of order and regularity into certain areas of the mass of signs. That is the role of relative motivation."[120]

Saussure differentiated between a lexicological and a grammatical pole between which the process of formation moved. Different languages exhibited different trends. Chinese, for example, embodied the lexicological pole, and Indo-Germanic languages and Sanskrit the grammatical. Concerning the production of grammaticality, he stated: When a single modification of sound entered general usage, alternation and analogy produced a series of similar sounding phenomena and in this way created subsequent regularity. Similar to Freudian rationalisation, these processes inscribed a spontaneous, non-purposive, yet collective mutation with an apparent regularisation through first creating examples. However, one cannot see how these phenomena might give rise to a more regular distribution of vowels.

This was accomplished by a self-evident limitation, which Saussure did not mention, that constrained sound changes from the outset. The few people who spontaneously began to speak in different ways were nevertheless constrained by the mouth, the organ of utterance. The semiologist Saussure mentioned this in his observations on phonetics:

"For it is not always within our power to pronounce as we had intended. Freedom to link sound types in succession is limited by the possibility of combining the right articulatory movements."[121]

Saussure demanded strict mathematisation, similar to that which Markov carried out a few years later:

"To account for what happens in these combinations, we need a science which treats combinations rather like algebraic equations. A binary group will imply a certain number of articulatory and auditory features imposing conditions upon each other, in such a way that when one of them varies there will be a necessary alteration of the others which can be calculated."[122]

This "chain of speech"[123] was bracketed together with the organs of speech:
"[A] normal continuous sequence [...] is characterised by a succession of graduated abductions and adductions, corresponding to the opening and closing of articulators in the mouth."[124]

---

120. Saussure, *Course*, p. 131.

121. Saussure, *Course*, p. 51.

122. Ibid. See also Benoit Mandelbrot, On the theory of word frequencies and on related Markovian models of discourse, in: *Structure of Language and Its Mathematical Aspects*, ed. Roman Jakobson (Rhode Island, 1961), pp. 190–219, p. 211: "To an extent unrivalled by other classics in the field, Saussure exhibited an 'esprit géométrique', which was most welcome to a mathematician."

123. Saussure, *Course*, p. 41.

124. Saussure, *Course*, p. 57.

The Vertical "Lalula" and Cryptography

To verify his findings concerning the connection between the observed dispersion and the dependence of the letters, Markov then proceeded to analyse a completely different arrangement of the text material in which the connection between the letters was destroyed. He turned his attention to the sums in the last columns of the main table, which gave the number of vowels per 100 letters, and which were attained when looking at every fifth letter of the 500 letters that each small table represented. The elements of this group were only weakly mutually dependent because they were far apart in the text. However, the five sums within one batch of five hundred were to a large degree mutually dependent because the individual letters stood in close proximity. The first contained the letters 1, 6, 11, 16, 21, 26, and so on; the second contained the letters that followed immediately after – 2, 7, 12, 17, 22, 27, etc. All 100 letters of the second were dependent on the first 100. Weaker connections existed between the tables. The first 100 of the second table contained only 50 letters (in column 1), which followed the last group of the first (those in column 9). The second arrangement thus gave 200 strongly dependent samples of 100 non-enchained letters. The number of vowels per 100 letters fluctuated between 26 and 57. The values were now much more dispersed; the table had to be split up because it was now too wide to fit on the page. The arithmetic mean was still 43.2 – the 20,000 letters examined were the same ones. The sum of the squares of deviations from this number was much higher: 5,788.8. By modifying an equation from his *Probability Theory*, Markov demonstrated that, contrary to the traditional view, the law of large numbers as well as the Gaussian method could also be applied to dependent quantities.[125] This expansion of stochastics to include cases that went against the dogma of independent events and were usually excluded for that reason was pursued by Markov in numerous papers throughout his life. Through dividing the above sum by the number of samples, 200, the mean variance of the numbers in the last columns was calculated to be 28.944. It was more than five times greater than in the first arrangement. Divided further by 100 (each value described 100 letters), the result was a 0.28944 mean dispersion of each single letter, which differed only slightly from the theoretical expectation of the binomial formula, $0.432 \times 0.568 = 0.245376$. The dispersion coefficient was 1.18: it was thus in Lexis' sense normally random and demonstrated that the single letters of the samples represented mutually independent events with the typical Gaussian random dispersion.[126]

---

125. A.A. Markov, *Probability Theory* (St. Petersburg, 1900). In Russian; in the German edition, A.A. Markov, *Wahrscheinlichkeitsrechnung* (Leipzig, Berlin, 1912), the equations used are found on pp. 203 and 209.

126. In the study of the Aksakov text, letters were combined with each other within the groups of ten that in the text were separated by nine letters. The coefficient normalised to 1.05, and thus corresponded again to a random distribution. Markov demonstrated this by making a table containing this count together with the theoretical distribution for 10,000 independent samples. For calculating the latter, he referred to a passage in his *Probability Theory*, which gave Newton's binomial formula; cf. Markov, *Wahrscheinlichkeitsrechnung*, p. 27.

A similar result would be expected if 200 batches of 100 balls were to be taken from an urn containing 4,319 white and 5,681 black balls.[127]

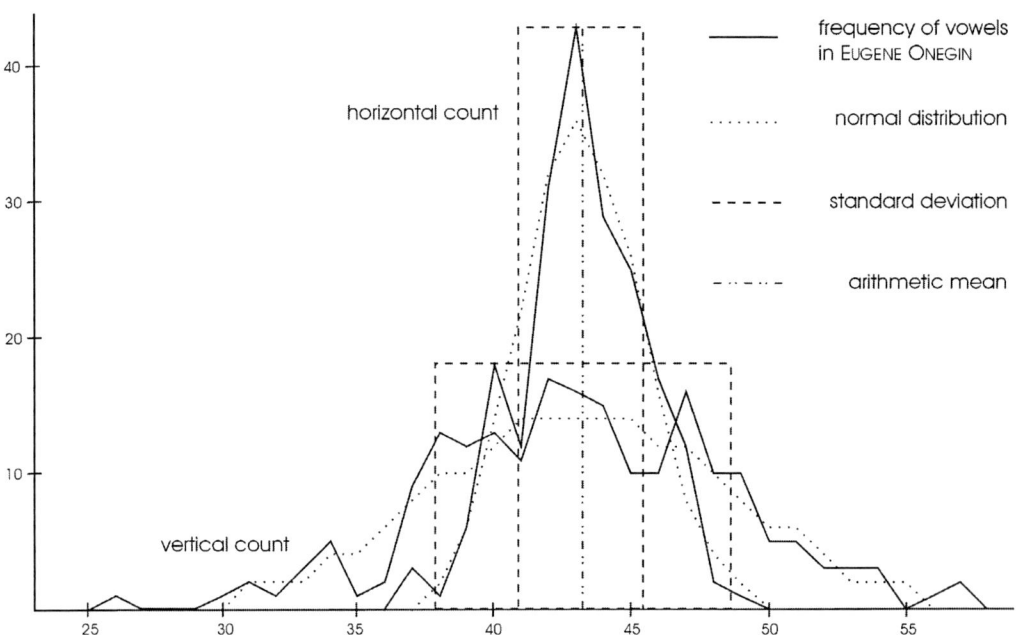

Fig. 5: The horizontal and vertical count compared.

Already in the count of the groups of 100 it was apparent that the direction in which one proceeded made a difference and had considerable consequences. The columns exhibited a greater deviation from the mean value than the rows. By constructing series of each tenth letter, the coherence of the words was lost. In the articulation of language, the necessity of making a vowel follow a sequence of consonants, and vice versa, did not apply here. What remained were disparate and discrete letters, devoid of meaning. This produced exceptional statistical phenomena, such as "myrnnschteme" in the first column with only

127. The calculation of the mathematical expectation of the square of error by dividing the variance again by the number of samples as above was not possible in the case of dependence of the quantities under consideration because the dispersion was artificially increased by the dependence and no longer corresponded to the fluctuation of the arithmetic mean that would be expected if another 20,000 letters of the text were to be examined. Markov therefore proposed using the result of the first arrangement for this, where the different values were not mutually dependent. The connection of the numbers was also clear from the fact that the sum of the squares of their deviations from the mean changed drastically when the values were combined in twos, fours, or fives. Instead of 5,788.8, the sums were 3,551.6, 3,089.2, and 1,004. Because of this combining of values, the groups contained progressively less letters, which were next to each other in the text. In the groups of 200, which were formed by combining in pairs, only half were mutually dependent, in the groups of 400, only a quarter, until in the groups of 500, there were none at all. The combination in fives, whether in columns or rows, each contained the values of one small table and produced entirely independent samples. For this reason, the result was the same for both: 1,004.

three vowels among ten letters or "ochaeeaawor" in the second with the amazing number of seven. These nonsensical word monsters could be taken from Morgenstern's *Lalula* of 1905 or the typographic poems of Raoul Hausmann after 1917: "klekwapufzi," or "fmsbwtözäu."[128]

Markov found the comparison material of independent samples, which allowed him to highlight the peculiarities of language, in one of the oldest known coding operations. Around 450 BC, in Lysander's Sparta, an instrument called a *scytale* was in usage that Plutarch described as follows:

> "When the ephors send an admiral or general on his way, they take two round pieces of wood, both exactly of a length and thickness, and cut even to one another; they keep one themselves, and the other they give to the person they send forth; and these pieces of wood they call *scytales*. When, therefore, they have occasion to communicate any secret or important matter, making a scroll of parchment long and narrow like a leathern thong, they roll it about their own staff of wood, leaving no space void between, but covering the surface of the staff with the scroll all over. When they have done this, they write what they please on the scroll, as it is wrapped about the staff; and when they have written, they take off the scroll, and send it to the general without the wood. He, when he has received it, can read nothing of the writing, because the words and letters are not connected, but all broken up; but taking his own staff, he winds the slip of the scroll about it, so that this folding, restoring all the parts into the same order that they were in before, and putting what comes first into connection with what follows, brings the whole consecutive contents to view round the outside. And this scroll is called a *staff*, after the name of the wood, as a thing measured is by the name of the measure."[129]

In cryptology, this type of encryption is called a transposition cipher because only the order of the letters changes, not the letters themselves as in substitution ciphers. The connections between the signs are what is erased. The variant used by Markov is known as columnar transposition; in 1881, the revolutionaries that murdered Czar Alexander II used a double version of this cipher. It was called "Nihilist transposition" after them and was "the most popular cipher of the Russian underground."[130] Modified versions of this cipher were in use until recent times, for example, the ÜBCHI cipher of the German Wehrmacht in the First World War or as a tool of many secret agents after the Second World War.[131] Even today, double columnar transposition is regarded as difficult to decipher.[132]

---

128. See Kittler, *Discourse Networks*, p. 212; also Bernhard Holeczek and Lida von Mengden, eds., *Zufall als Prinzip* (Heidelberg, 1992), p. 77.

129. Plutarch, *The Lives of the Noble Grecians and Romans* [ca. 75 AD] (Chicago et al., 1952), p. 362 (Lysander).

130. David Kahn, *The Codebreakers. The Story of Secret Writing* (New York, 1967), p. 619.

131. See Otto Leiberich, Vom diplomatischen Code zur Falltürfunktion. *Spektrum der Wissenschaft. Dossier Kryptographie* (2001): 12–18, p. 17: "In the early 1960s, a foreign espionage organisation equipped their agents operating in the Federal Republic of Germany with an encryption technique, double columnar transposition, that was regarded as secure. The foreign cryptologists, however, made one mistake. They gave their agents mnemonic sentences, which they had taken from literary works, as keys." (My translation – D.L.)

132. Cf. Wayne G. Barker, Cryptanalysis of the Double Transposition Cipher (Laguna Hills, 1995).

It is safe to assume that Markov knew of this cipher and that it provided inspiration for his study.

Cryptography of this kind also played a role in the history of the poem *Eugene Onegin*. In 1904, the encrypted fragments of a tenth chapter were presented to the autograph section of the St. Petersburg Academy of Sciences. The literary historian Piotr Osipovich Morozov, who became a corresponding member of the Academy in 1912, deciphered Pushkin's text three years before Markov's lecture. He published the plaintext in 1910 in the Academy journal *Puskin i ego sovremenniki*. It was a harsh critique of the Czar and a description of the Decabrists' circles. Pushkin, who was already persecuted on political grounds, judged the content so explosive that he destroyed the manuscript on 19 October 1830, five years after the attempted *coup d'état*. A short time later, he re-wrote the chapter from memory, but this time in code.

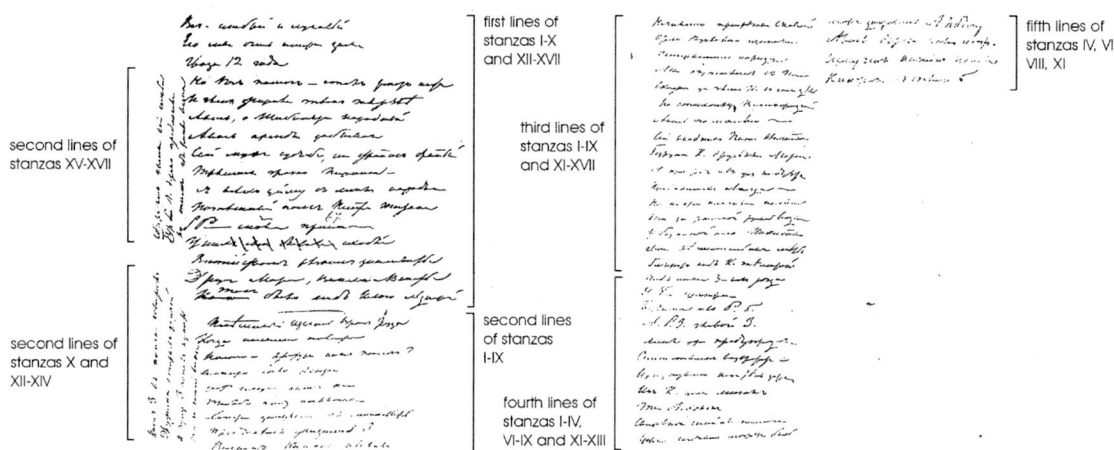

Fig. 6: Pushkin's cryptogram of the tenth chapter.[133]

It is most likely that Markov knew about the sensational discovery and deciphering of Pushkin's tenth chapter. Nabokov gave this description of the cryptogram:

"A column of sixteen lines [...] representing the first lines of sts. I–X and XII–XVII. Under this, separated by a horizontal dash, another column [...] representing the second lines of sts. I–IX. Two sets of lines [...] in the left-hand margin, the lines parallel to the margin; the lower marginal set represents the second lines of sts. X, XII–XIV, and the upper marginal set represents the second lines of sts. XV–XVII. A column of twenty-seven lines [...] down the left-hand side of the page, representing the third lines of sts. I–IX and XI–XVII, followed (without any gap or dash) by the fourth lines of sts. I–IV, VI–IX, and XI–XIII.

133. Boris Tomashevski, The 10th chapter of *Eugene Onegin*. The story of its solution. *Literaturnoe nasledstwo* 16–18 (1934): 378–420, p. 384f. In Russian. Cf. Piotr Morozov, Pushkin's coded poem. *Puskin i ego sovremenniki* 4 (1910): 1–12. In Russian.

A column of four lines [...] at the top of the right-hand side of the paper, representing what I take to be the fifth lines of sts. IV, VI, VIII, XI."[134]

Pushkin combined lines of the poem, which occupied the same positions in the stanzas, to conceal their context and meaning. He arranged whole paragraphs in lines underneath each other, in which each verse had its own column, and then read them vertically. He attempted to disguise particularly disparaging passages further by using abbreviations: "O R[ússkiy] glúp[ïy] násh na[ród]" – "O our R[ussian] stup[id] na[tion]."[135] In Nabokov's opinion, he capitulated after encoding the fifth verse: "I also suggest that he soon noticed that something was very wrong with his cipher and in utter disgust gave up the whole matter."[136] The method of the mathematician Markov, which combined letters that were separated in the text by four letters, thus precisely mirrored Pushkin's abandoned attempt to encipher his poem, where verses with the same distance from each other were brought together. However, that Markov combined the vertical counts of ten letters each so that each fifth letter was selected, also had a different reason. This compression was necessary because otherwise the results for 20,000 letters would not have fit onto one page. On the other hand, stronger compression would have resulted in part in three-figure sums. This did not affect the mathematical results because the combination of letters, which were separated by four others, was almost as weak as if separated by nine.

Traces of the Mouth

Markov's analysis found something within text that writing supersedes, in the two-fold sense of destroying and preserving: orality, more precisely, in the form of its physical organ – the mouth. Traces from the speech organ influence written text in two ways. On the one hand, it is free to mutate individual sounds and see if others will accept these mutations. On the other hand, it is constrained in general by the fact that it can only articulate certain combinations. This constraint becomes obvious when that which is formed in speech is discretely fixed in written text. On the two-dimensional page of printed letters, the potential for re-combination, and thus the reservoir of chains of letters that are possible, but not in use, becomes strikingly obvious. This knowledge can then be used to conceal messages by transposition ciphers, as Pushkin did with the tenth chapter of *Eugene Onegin*.

Markov proved that language is better grasped when one includes the relationships between its elements. Saussure formulated this in a more radical way:

"*In the language itself, there are only differences.* Even more important than that is the fact that, although in general a difference presupposes positive terms

---

134. Pushkin, *Eugene Onegin*, p. 365f.
135. Pushkin, *Eugene Onegin*, p. 370.
136. Pushkin, *Eugene Onegin*, p. 374.

between which the difference holds, in a language there are only differences, *and no positive terms.*"[137]

For the first time in mathematics the use of signs was treated as being differential. The single letters were not considered to be important, as in early cryptanalysis, rather it was the connections between them. In this way, d'Alembert's problem, mentioned at the beginning of this chapter, was at least partially solved. Transition probabilities now allow us to calculate that the word "constantinopolitanensibus" uttered by a person is more probable than "nbsaeptolnoiauostnisnictn." Although both contain the same letters, that which is unsayable is sorted out. This method, therefore, also determines the degree to which text represents orality.

---

137. Saussure, *Course*, p. 118.

There Must Be an Angel
On the Beginnings of the Arithmetics of Rays

From August 1953 to May 1954 strange love-letters appeared on the notice board of Manchester University's Computer Department:[138]

DARLING SWEETHEART

YOU ARE MY AVID FELLOW FEELING. MY AFFECTION CURIOUSLY CLINGS TO YOUR PASSIONATE WISH. MY LIKING YEARNS FOR YOUR HEART. YOU ARE MY WISTFUL SYMPATHY: MY TENDER LIKING.

YOURS BEAUTIFULLY

M. U. C.[139]

The acronym "M.U.C." stood for "Manchester University Computer," the earliest electronic, programmable, and universal calculating machine worldwide; the fully functional prototype was completed in June 1948.[140] One of the very first software developers, Christopher Strachey (1916–1975), had

---

138. T. William Olle, personal communication, 21 February 2006: "I do remember a copy of the Strachey love-letter being put on the notice board and that must have been after August 1953 and probably prior to May 1954 (the date of Alan Turing's death)."

139. Christopher Strachey, The "thinking" machine. *Encounter. Literature, Arts, Politics* 13 (1954): 25–31, p. 26.

140. Frederic C. Williams and Tom Kilburn, Electronic digital computers. *Nature* 162 (1948): 487. By "computing" and "calculating" I mean here and in the following in general the processing of data. Various machines are claimed to be the "first" computer, but all others lack one of the properties mentioned. The "ABC," developed by John V. Atanasoff and Clifford Berry 1937–1941 in the USA, was a binary digital "equation solver" and remained unfinished due to World War II intervening. From 1938, in Berlin Konrad Zuse constructed a series of electro-mechanical binary digital equation solvers, culminating in 1941 in the functioning model "Z3." Both projects included a certain internal memory for numbers but not for instructions. The same applies to "COLOSSUS," completed in the UK in December 1943, and the North American "ENIAC" of November 1945. On both machines the instructions were wired on a plugboard. See Simon Lavington, *Early British Computers* (Manchester, 1980), p. 4ff. Lavington offers a very easy and readable account of the early history of computers in Great Britain.

used the built-in random generator of the Ferranti Mark I, the first industrially produced computer of this kind, to generate texts that are intended to express and arouse emotions. The British physicist performed this experiment a full thirteen years before the appearance of Joseph Weizenbaum's *ELIZA*, which is commonly – and mistakenly – held to be the earliest example of computer-generated texts.[141]

Using numerous resources found on the Internet, I constructed an emulator of the Ferranti Mark I, and ran Strachey's original programme on it, which is preserved in his papers held by the Bodleian Library in Oxford.[142] Thus the following analysis of how the hard- and software functions is not only based on theoretical consideration of the subject matter, but equally on transforming "thought into being and put its trust in the absolute difference" during the long and arduous reconstruction of the details, and to "stretching it [the mind] on the rack in order to perfect it as a machine."[143]

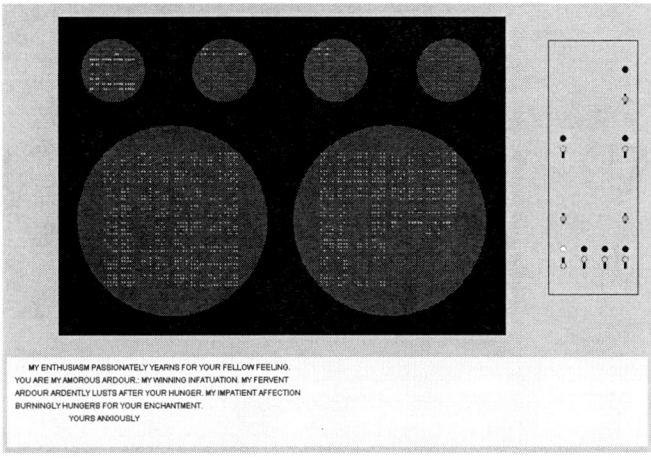

Fig. 1: The user interface of the Ferranti Mark I emulator with the Love-letters algorithm loaded.

141. Joseph Weizenbaum, ELIZA. A computer program for the study of natural language communication between man and machine. *Communications of the ACM* 9 (1966): 36–45.

142. I particularly used documents that Brian Napper has made available on the excellent website http://www.computer50.org/, Alan Turing's *Programmers' Handbook for the Manchester Electronic Computer Mark II* (typescript, Manchester, 1951), available online in two versions, http://www.turingarchive.org/browse.php/B/32, http://www.alanturing.net/turing_archive/archive/m/m01/m01.php, transcript at http://curation.cs.manchester.ac.uk/computer50/www.computer50.org/kgill/mark1/progman.html, and Dietrich G. Prinz, *Introduction to Programming on the Manchester Electronic Digital Computer* (typescript, Manchester, 1952), online: http://www.alanturing.net/turing_archive/archive/m/m11/M11-001.html. The manuscripts of the love-letter programme are held in the Special Collections and Western Manuscripts section of the Bodleian Library, Oxford University, CSAC 71.1.80/C.34 and C.35. My warmest thanks go to Brian Napper, Christopher Burton, Martin Campbell-Kelly, Simon Lavington, and T. William Olle for their exceptionally generous support, tips, and suggestions.

143. Georg W.F. Hegel, *The Phenomenology of Spirit* [1807], trans. A.V. Miller (Oxford, 1977), p. 400 (the "beautiful soul"); idem, *Science of Logic* [1812], trans. A.V. Miller (London, 1969), p. 217.

Programme of a Love-letter

After studying mathematics and physics at King's College, Cambridge, during the war Christopher Strachey worked for Standard Telephones and Cables Ltd. in London on electron tubes for centimetric radar. In this work he made use of the differential analyser invented by Vannevar Bush, which awakened his interest in computers.[144] After the capitulation of Germany, Strachey became a schoolteacher. In January 1951 a friend introduced him to Mike Woodger of the National Physical Laboratory (NPL). The lab had successfully built a reduced version of Turing's automatic calculating engine (ACE), a concept that dated from 1945: the Pilot ACE. In May 1950 the first computations were performed on this machine. After the meeting with Woodger, in his spare time Strachey developed a programme for the game of draughts, which he finished in February 1951. The game completely exhausted the Pilot ACE's memory. The draughts programme ran for the first time on 30 July 1951 at NPL, and developed into an early attempt at getting a computer to write its own programme, so-called auto-coding.

When Strachey heard about the Manchester Mark I, which had a much bigger memory, he asked his former fellow-student Alan Turing for the manual, transcribed his programme into the operation codes of that machine by around October 1951, and was given permission to run it on the computer.

> "Strachey sent his programme for punching beforehand. The programme was about 20 pages long (over a thousand instructions), and the naiveté of a first-time user attempting a programme of such length caused not a little amusement among the programmers in the laboratory. Anyway, the day came and Strachey loaded his programme into the Mark I. After a couple of errors were fixed, the programme ran straight through and finished by playing 'God Save the King' on the 'hooter' (loudspeaker). On that day Strachey acquired a formidable reputation as a programmer that he never lost."[145]

Because of this achievement, the National Research and Development Corporation (NRDC) offered Strachey the post of technical officer the following month. Figure 2 shows the general structure of the Love-letters software, in Strachey's handwriting, that he developed in June 1952 along with two other small projects soon after joining NRDC.[146]

---

144. On Strachey's biography see Martin Campbell-Kelly, Christopher Strachey, 1916–1975. A biographical note. *Annals of the History of Computing* 7 (1985): 19–42.

145. M. Campbell-Kelly, Programming the Mark I. Early programming activity at the University of Manchester. *Annals of the History of Computing* 2 (1980): 130–168, p. 133. This excellent article describes the machine in detail. Strachey's accomplishment is all the more admirable since Turing's manual teems with mistakes and inaccuracies. It forced the reader to do some "de-bugging" immediately and painfully to complete a learning process of the foreign language. Cf. Frank Sumner, Memories of the Manchester Mark 1. *Computer Resurrection. The Bulletin of the Computer Conservation Society* 10 (1994), 9–13, p. 9: "So all the programmes written in Mark 1 code had slight errors in them, and by the time you had worked out what the code should have been you had become quite a competent programmer."

146. The other two projects, commissioned by NRDC, concerned the computation of a surge shaft and the Ising model of ferromagnetism; cf. National Cataloguing Unit for the Archives of Contemporary Scientists

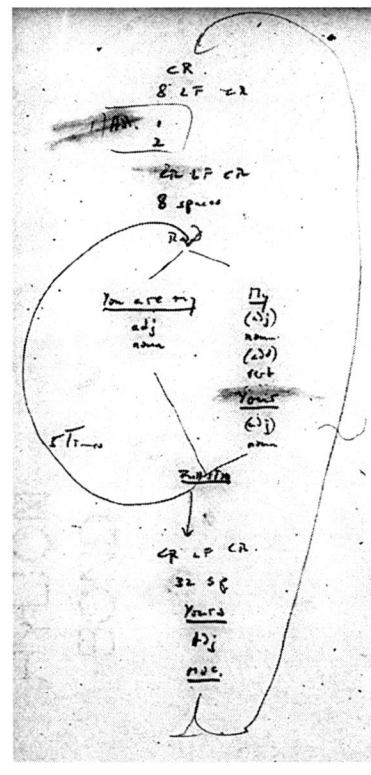

Fig. 2: Schematic of the Love-letters programme[147]

Apart from position commands like carriage return ("CR"), line feed ("LF"), and spaces ("spaces" or "sp"), the algorithm prints two salutations ("Add." = address). Then it enters a loop, which is carried out "5 times" and, depending on a random variable ("Rand"), follows one of two alternative paths. One generates a sentence following the syntactic skeleton "You are my – Adjective (adj) – Substantive (noun)"; the other path gives "My – [Adjective] – Substantive – [Adverb (adv)] – Verb (verb) – Your – [Adjective] – Substantive" (the static words are underlined, the optional words are in square brackets). The first sentence of the example given at the beginning of this chapter follows the first scheme, and the second sentence follows the other. Each phrase ends with a "Full stop." After the programme leaves the loop, it closes with the ending "Yours – Adverb (in the schematic this is given erroneously as "Adj") – MUC."

The University of Manchester's Computer

From a technical perspective, the Universal Machine that Alan Turing (1912–1954) designed theoretically in 1936 can be reduced to a problem of memory.[148] It had to be capable of writing, reading, storing, and deleting any data. To this end, the engineer Frederic Williams (1911–1977) had modified cathode ray tubes common in both warfare and commercial applications (CRTs – like the ones that were once used in television sets) for the Manchester Mark I in such a way that electronics repeatedly read and refreshed the 1280 picture dots.[149]

(NCUACS), Catalogue of the Papers and Correspondence of Christopher Strachey (1916–1975). CSAC no. 71/1/80 (Bath, 1980). http://www.a2a.org.uk/search/extended.asp, Catalogue ref. NCUACS 71.1.80.

147. All images of the original programme are from the Special Collections and Western Manuscripts section of the Bodleian Library, Oxford University, CSAC 71.1.80/C.34 and C.35 by kind permission.

148. A.M. Turing, On computable numbers, with an application to the Entscheidungsproblem. *Proceedings of the London Mathematical Society (Ser. 2)* 42 (1937): 230–265. Cf. F.C. Williams and T. Kilburn, A storage system for use with binary-digital computing machines. *Proceedings of the Institution of Electrical Engineers, Pt. 2* 96 (1949): 183–202, p. 183: "The problem of electronic digital computing from the engineering standpoint, lies primarily in the construction of suitable electronic devices having the same number of states as the number of possible values of a digit."

149. Williams also began his career in radar research, which is discussed below.

The screen was divided into two columns each with 32 lines of 20 bits. Eight of these monitors were employed to load and run programme pages (data and operations). The size of the main working memory was around 1.25 kilobytes. On two monitors connected in parallel the user could directly see the content of the various tubes.[150]

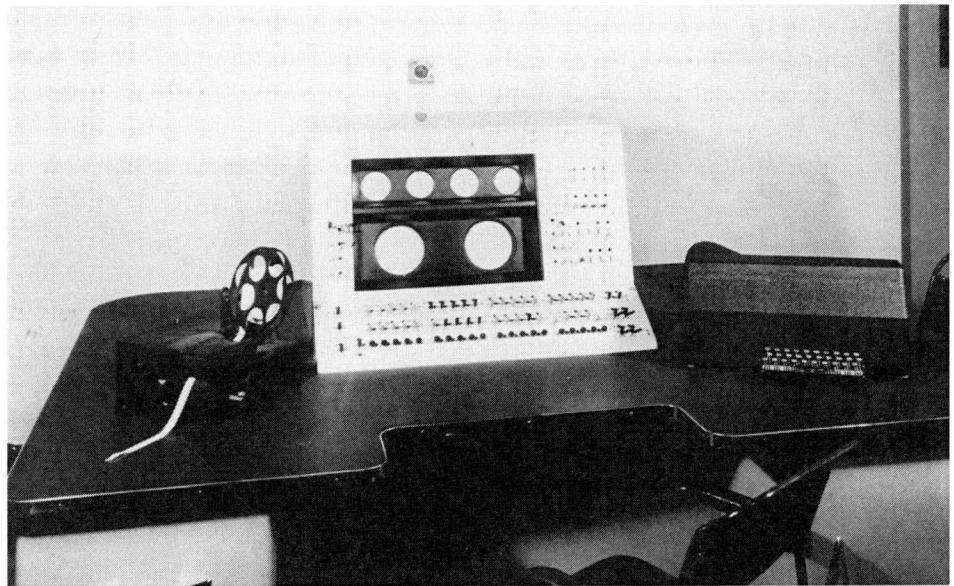

Fig. 3: The console of the Ferranti Mark I. Top: The monitors B, C, A, and D. Bottom: Two pages of the main memory.[151]

A magnetic drum was available for the long-term storage of data. The programmer could load information from there onto one of the monitors or save information from the monitor in the drum.[152] The actual processing was done on four smaller tubes, whose content was also visible, and which were labelled with the first four letters of the alphabet. A, the accumulator, contained the results of the

150. These are the two larger circular disks in Figure 3. Cf. Campbell-Kelly, Mark I, p. 154: "The ease with which monitor tubes could be used with CRT storage was a very attractive feature of this technology; there is no equivalent for today's memories. By sitting at the console, the programmer could observe the progress of the programme on the monitor tubes in a process known as 'peeping.' Peeping was very much the modus operandi at Manchester."

151. Bertram V. Bowden, ed., *Faster than Thought. A Symposium on Digital Computing Machines* (London, 1953), p. 127.

152. The magnetic drum was located in a room above the actual computer workshop, which led to the introduction of the terms "down transfer" and "up transfer" for these two operations that live on in the modern variants "upload" and "download." Cf. Prinz, *Introduction*, p. 23; F.C. Williams, Early computers at Manchester University. *The Radio and Electric Engineer* 45 (1975): 327–331, p. 328: "The two-level storage I have referred to was indeed on two levels. The electronic store was in the magnetism room and the magnetic store in the room above. Transfers between the stores were achieved by setting switches, then running to the bottom of the stairs and shouting 'We are ready to receive track 17 on tube 1.'"

arithmetical and logical operations and also temporarily stored data for the transmission from one line of the page to another. In the C tube (C for control) was the current instruction and its address. The most momentous invention for the later development of computers lay in the auxiliary store B, which was given this letter because A and C were already in use. The content of B could be added to the current command and thus could modify it before it was carried out. Today, this is termed "index addressing" and it allows a single instruction to be applied to a list of any length.[153] Finally, D contained the multiplier in appropriate calculations. The computer not only displayed the data, operations, and addresses on the CRTs (this must be considered a side-effect of the chosen medium), it stored, read, wrote, and processed these in the electrical charges of the picture dots.[154]

Routines of Love

Strachey's programme filled four double monitor storage units. The word data needed for the letters was located on the last three pages, written backwards; the main algorithm was on the first page.[155]

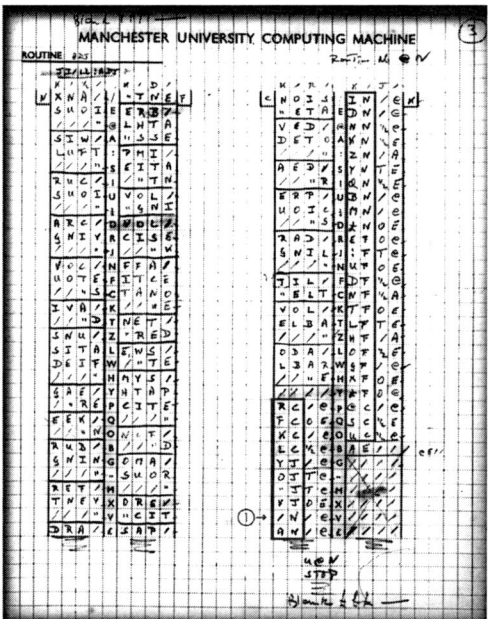

Fig. 4: The list of adjectives in Love-letters, page 3 of the programme.

---

153. Cf. Campbell-Kelly, Mark I, p. 135.

154. As we have seen in the first chapter, this positions the symbols, which the machine processes, before differentiation between letter and number. The symbols simply mark the pure difference.

155. See Figure 4. The first adjectives are: "anxious," "wistful," "curious," and "craving."

In addition the software had two other parts: "PERM" and "ENGPRINT." "PERM" enabled sub-routines to be linked in and belonged as it were to the "operating system," the Scheme A. It overwrote the main programme in memory with another, executed it, then restored the original state, and followed the instructions from the point at which it had left them. "ENGPRINT," the only sub-algorithm the programme used, printed the symbols at the address, which was set in Line 4 of Tube B; it printed them line by line and backwards until it encountered the meta-symbol """. To generate text from the words stored on pages 2–4, the variable only had to be assigned a value repeatedly and the print routine called up. The problem of generating text was reduced to the administration of addresses.

A "word" of the Mark I – the smallest unit of information used – was 20 bits long. On the recommendation of Geoff Tootill, and later Turing, outside the computer these were represented by four letters of the Baudot alphabet (common in telegraphy), each of which was 5 bits. The sequence of symbols is shown in the middle columns of Figure 4, where they number the lines of the programme: "/E@A:SIU...." As one can see in the boxes that are appended at the top of both sides of the tables, these are Columns 13–16 (N, F, C, K), which constituted the fourth of the eight tubes. Simon Lavington aptly notes: "To Turing, who had spent countless hours at Bletchley Park battling with Geheimschreiber 5-bit ciphers during the war, the teleprinter code must have seemed very natural. To lesser mortals it was painful!"[156] Working with this representation was made even more difficult by the fact that the usual order of numbers was reversed. The so-called "highest significant bit" (the bit with the highest value) was not on the left but on the right; 1000 was thus represented as 0001. Campbell-Kelly explains: "The reason for this is that in a serial machine, the digits are produced least-significant first; Williams tubes and oscilloscopes conventionally sweep with time going from left to right, so it was natural to write binary numbers that way and this was common on early computers."[157] The addresses of the lines were also written backwards in this notation. The word "anxious" at the top left of Figure 4 is in position "/N," which together with the positions of all the other words on the page is found in the right-hand table, in the first two signs of the penultimate line of the left-hand column (1). "AN" beneath it references "wistful," etc.

156. Lavington, *Early British Computers*, p. 42.
157. Campbell-Kelly, Mark I, p. 136.

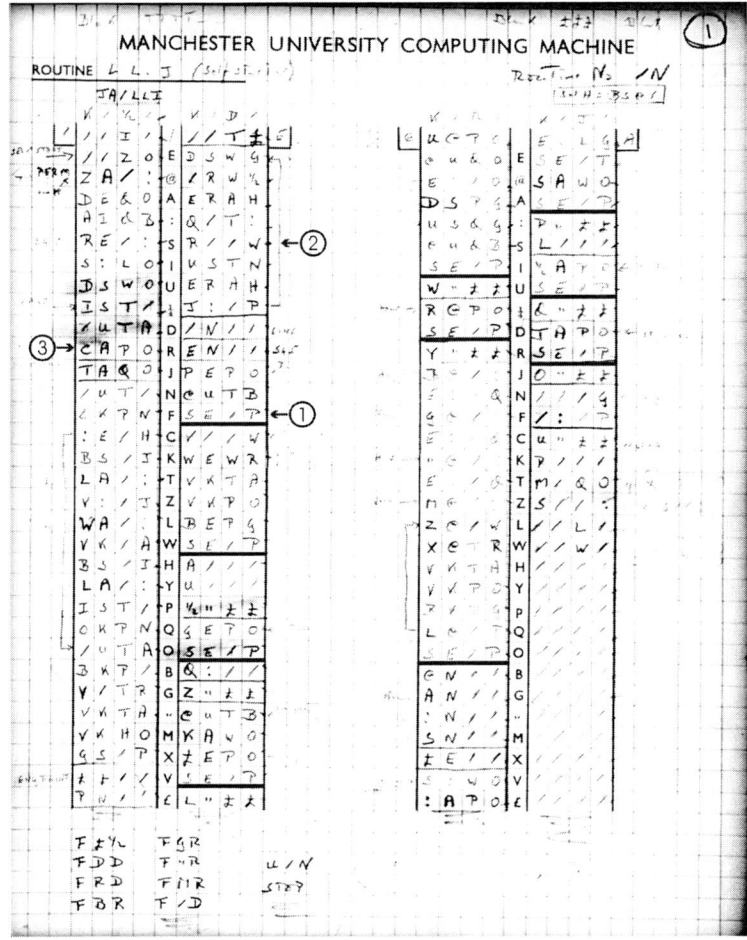

Fig. 5: The first page of Strachey's algorithm.

To understand the structure of a command of the Manchester computer, let us take as an example the sequence "SE/P," which appears in Line "FE" of the main programme, amongst others (1 in Figure 5). In general, in the machine's instruction set the last two symbols indicate the command that is to be performed, and the first two symbols indicate the address to which it refers. "/P" means an unconditional jump to the line that "SE" references. At position "SE" the algorithm finds the first two signs "R/" (2). Thus "SE/P" encodes the instruction to continue with execution in Column 1, Line 11 (3). Through this command that stands at the end of many of the sections that are separated by horizontal lines, the software always returns to its beginning in order to execute one of the routines for the definition of the address of the next word to be printed – dependent on Line 3 of Tube B, to write the address in Line 4, and to call up the output-programme.

Variable Scripts

Astonishingly, according to Joseph Weizenbaum, he knew nothing of Strachey's experiments when he wrote *ELIZA*, although in this period computer departments in England and North America maintained lively communications and visited each other regularly.[158] This is even more odd considering that in 1962, Strachey corresponded with Weizenbaum, and at the time *ELIZA* was published, Strachey was Guest Lecturer at the Massachusetts Institute of Technology (MIT) for seven months. Strachey had also cleared up the organisational issues concerning his term at MIT with Weizenbaum. The British scientist continued to give summer lectures regularly at MIT until 1970.[159] I, too, knew nothing about Strachey's programme when I wrote my Ph.D. dissertation on early algorithms for generating text.[160] In my endeavour to derive the more complex procedures from the simpler ones, I selected an example dating from 1997, because I could not find a study object to illustrate the most basic form of meaning production, which logically preceded *ELIZA* and was chronologically fitting.[161] Strachey's programme discussed here closes the gap in a satisfying way and in general falls into the category of variable scripts. A series of abstract signifiers references a list of equally possible, concrete instantiations, which are inserted for them at random and independent of each other. The fascinating thing about this was and is that seemingly endless variety can be generated from a relatively small group of words, as expressed in the title of one of the first literary experiments in this direction, Raymond Queneau's *Cent Mille Milliards de Poèmes*.[162] Strachey strengthened the impression of diversity by not only varying the words used, but also including optional elements, which were sometimes omitted and thus modified the structure of the sentences. Additionally, two different syntactic structures alternated at random. When the construction "You are my – Adjective – Substantive" repeated, the programme shortened the second instance to ": my – Adjective – Substantive," thus cleverly avoiding repetition.[163] In total, Strachey's software could generate over 318 billion different love-letters.

Like *ELIZA*, *Love-letters* used personal pronouns to create a relationship between two communication partners. Both sentence constructions used relate "my" to "you," or "your," but not in the form of a dialogue where "you" would be transformed on the other side into "me" and vice versa, as is the case with *ELIZA*. Because *Love-letters* did not display the result but printed it because this was easier to realise technically, the addressee of the letters remains ambiguous. The computer is either writing to or for its user. Ultimately, the software is based on

---

158. Joseph Weizenbaum, personal communication, 3 May 2006.

159. Cf. NCUACS, *Catalogue of Strachey Papers*, CSAC 71.1.80/A.66, CSAC 71.1.80/C.202–C.207.

160. David Link, *Poesiemaschinen / Maschinenpoesie. Zur Frühgeschichte computerisierter Texterzeugung und generativer Systeme* (Munich, 2006; in German).

161. Nick Sullivan, Romance Writer. Computer-Generated Romance Stories [1997]. http://www-ssrl.slac.stanford.edu/~winston/baers/romriter.html.

162. Raymond Queneau, *Cent Mille Milliards de Poèmes* (Paris, 1961).

163. Cf. the end of the example quoted at the beginning of this chapter.

a reductionist position vis à vis love and its expression. Like the draughts game that Strachey had attempted to implement the previous year, love is regarded as a recombinatory procedure with recurring elements, which can be formalised, but which is still intelligent enough to raise considerable interest should the simulation succeed.

The Quest for a Magic Writing-Pad

On the one hand, the conditions that make computers possible are acquired by the reduction of fundamental arithmetic operations into the simple transformation of primitive symbols according to rules, which themselves can be represented and treated as symbols. From 1906 onward, the Norwegian mathematician Axel Thue initiated work on this reduction; Kurt Gödel, and then Alan Turing, developed it together with all the paradoxes that resulted.[164] However, this kind of understanding of symbols has implicitly existed in cryptography since Girolamo Cardano (1550/1561) and Blaise de Vigenère (1586); today the term is "autokey." Like calculation with bits, secret ciphers transform a chain of source symbols (the plain text) using fixed rules (the key, often in the form of a tableau) into a different text (the cipher text). What is special about the autokey is the fact that the substitution routine directly depends on the message that is to be communicated. Thus already here data and operations converged insofar as both are symbols.[165]

On the other hand, and very practically, the symbols also had to be represented in their fluidity. This demanded a medium that was capable of both storing and selectively "forgetting." In 1925, when he was nearly seventy, Sigmund Freud wrote in his *Note upon the Magic Writing-Pad*:

> "All the forms of auxiliary apparatus which we have invented for the improvement or intensification of our sensory functions are built on the same model as the sense organs themselves. [...] Measured by this standard, devices to aid our memory seem particularly imperfect, since our mental apparatus accomplishes precisely what they cannot: it has an unlimited receptive capacity for new perceptions and nevertheless lays down permanent [...] memory-traces of them."[166]

In a close parallel to Turing's experience when constructing his Universal machine, Freud turns to the medium that he uses for writing notes: the sheet of paper. This material offers him the possibility of jotting down his thoughts, but not

---

164. Axel Thue, Über unendliche Zeichenreihen. *Kristiania Vidensk. Selsk. Skrifter. I. Mat. Nat. Kl.* 7 (1906): 1–22; Kurt Gödel, Über formal unentscheidbare Sätze der *Principia Mathematica* und verwandter Systeme, I. *Monatshefte für Mathematik und Physik* 38 (1931): 173–198; Turing, Computable numbers.

165. Cf. David Kahn, *The Codebreakers. The Story of Secret Writing* (New York, 1967), p. 143ff. Here I can only mention this connection in passing.

166. Sigmund Freud, A note upon the "Mystic Writing-Pad" [1925], in: *The Standard Edition of the Complete Psychological Works of Sigmund Freud*, ed. James Strachey (London, 1961), vol. 19, pp. 225–232, quotation p. 228. The editor and psychoanalyst James Strachey was Christopher Strachey's uncle.

to delete or change the symbols, except if he uses an eraser. As with Turing, the mistrust of the memory or forgetfulness makes it plausible to export the mental functions entirely.[167] Because Freud is unable to erase the symbols already written down, he is obliged to continue recording his thoughts on a new page when the first is full. He can turn back through the notes, which are possibly interdependent, to re-read the contents. The process of writing, however, is step-wise and only moves forward, limited like a Markov chain.[168]

Freud discovers in this magical toy a procedure that records in a radically different way. A single area is filled with symbols and then erased, over and over again. "If we imagine one hand writing upon the surface of the Magic Writing-Pad while another periodically raises its covering-sheet from the wax slab, we shall have a concrete representation of the way in which I tried to picture the functioning of the perceptual apparatus of our mind."[169] In 1923, a certain Howard L. Fischer from the Brown and Bigelow Co., Minnesota, USA, applied for a patent for this device, which was named the "Perpetual Memorandum Pad."[170] On the surface of the lower, wax-coated leaf, which the design inherits from Aristotle, the symbols overlap in natural stochastics to form traces.[171] In this device, the letter currently visible is only a condensed interim result of all the writing that has been done before. It loses its solidity. Future note-taking may strengthen other traces and the constant succession of setting down and erasing, by lifting up the top transparent leaf, may produce a different result. As at the beginning of Hegel's *Logic*, the alternation of "being" and "nothing" on the top page of the magic pad produces the possibility of "becoming" on the second underlying page: here the symbols transform.[172] Freud's interpretation, if not the invention of the magic pad, is clearly stamped with the concept of cinematography. Constancy and duration arise through, and are re-interpreted as, constant repetition of the same, as repetition of identical acts of writing.

---

167. Cf. Turing, Computable numbers, p. 253: "It is always possible for the [human] computer to break off from his work, to go away and forget all about it, and later to come back and go on with it." and Freud, Writing-pad, p. 227: "If I distrust my memory – neurotics, as we know, do so to a remarkable extent, but normal people have every reason for doing so as well – I am able to supplement and guarantee its working by making a note in writing."

168. Cf. Chapter 2.

169. Freud, Writing-Pad, p. 232, translation slightly altered. Freud is already thinking in terms of a periodic cycle.

170. Cf. the DEPATIS system of the German Patent Office (Deutsches Patent- und Markenamt, DPMA); on-line: http://depatisnet.dpma.de/DepatisNet/depatisnet, Patent no. US 1,543,430.

171. Cf. Aristotle, On the soul, in: *The Works of Aristotle. Vol. 3. Meteorologica, De mundo, De anima, Parva naturalia, De spiritu*, ed. William D. Ross and John A. Smith (Oxford, 1908–1952), Book 2, Chapter 12: "By a 'sense' is meant what has the power of receiving into itself the sensible forms of things without the matter. This must be conceived of as taking place in the way in which a piece of wax takes on the impress of a signet-ring without the iron or gold."

172. Cf. the first chapter.

Electrical Echoes

Also the "delay line," one of the first inventions for the volatile storage of data, achieved this sameness by repetition.[173] A piezo-electric crystal transformed electric oscillations into ultrasound waves, which excited water or kerosene in a tube; in later versions of the apparatus mercury was used. The waves travelled through the tube, and were then taken up by another quartz crystal, which amplified the waves and fed them back into the front of the tube. The writing, the time-delayed reading, and the re-writing of what was read effectuated continuity. The fundamental cycle, or loop, implemented the duration of symbols under technically changed conditions.[174] The delay line was not invented after the Second World War by Presper Eckert in North America, as is often maintained, but already in 1938 by William S. Percival at Electric and Musical Industries Ltd., EMI for short, at Hayes, in Middlesex, England, in connection with work on reducing interference in the transmission of moving images. In 1929 an extremely creative team of engineers formed at EMI that was headed by Isaac Shoenberg and Alan Blumlein. Their inventions included fundamental advances such as stereophony (1931) and electronic High Definition Television (from 1933).[175] The latter culminated in 1936 in the introduction of HDTV for the first BBC transmitting station at Alexandra Palace in London. Percival's system used an auxiliary channel, which only transmitted the interference noise encountered, with the purpose of interrupting the main signal when a particular threshold was crossed and replacing it with neutral data like a grey value. To gain time for processing and generating the control signal, it was necessary to delay the stream of images.[176]

---

173. Before the invention of the delay line other elements also existed for temporarily preserving "states," such as electro-mechanical relays or flip-flop circuits from Braun tubes; however, these could not be used extensively due to small storage capacity, relative expense, and slow switching. Cf. Lavington, *Early British Computers*, p. 13ff.

174. Freud had already come across a medium that could be used as volatile storage in one of his earliest texts, the *Project for a Scientific Psychology* from 1895, where he wrote about the "striking contrast" of the properties of nervous tissue and "the behaviour of a material that permits the passage of a wave-movement and thereafter returns to its former condition." The only idea missing was the loss-less and, therefore, infinite echo. Cf. Alexandre Métraux, Metamorphosen der Hirnwissenschaft. Warum Freuds "Entwurf einer Psychologie" aufgegeben wurde, in: *Ecce Cortex. Beiträge zur Geschichte des modernen Gehirns*, ed. Michael Hagner (Göttingen, 1999), pp. 75–109, p. 102.

175. For the early history of television, see Siegfried Zielinski, *Deep Time of the Media*, trans. Gloria Custance (Cambridge, MA, 2006), p. 236f. Zielinski attributes the development of the technology to the St. Petersburg electrochemist and -physicist Boris L. Rosing; Shoenberg studied with Rosing before the 1918 Revolution.

176. Cf. DEPATIS, Patent no. US 2,263,902 and US 468,994. Televisions still contain delay circuits today.

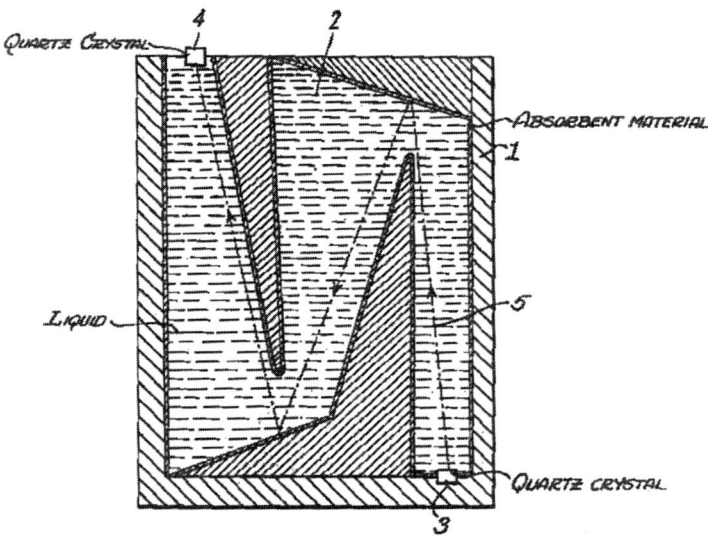

Fig. 6: Diagram of the "Delay Device" in William Percival's patent.[177]

With live transmission of television to a vast number of receivers, the recording and reproduction procedures, in which the "pencil of Nature" determined the representation, reached a limit.[178] Previously, media such as photography had claimed to deliver a faithful, true-to-life image of nature where no subjective will or style intervened. Now various kinds of interference that were equally natural threatened to severely disturb the distribution channels.[179] Whereas in the cinema it was possible to exert tight control over the distribution of information, TV technicians were confronted by a bewildering array of diverse scenarios for transmitting and receiving images as well as various external influences on them.[180] Percival's idea represents a first, simple solution of the problem that Nature often writes itself a bit too much into any recording un-mediated by a subject. The filtering of the data contains an automatic, numerical comparative operation and produces a blind area in the stream of moving representations.

The delay line makes use of the fact that, in send/receive systems like telegraphy or telephony, information requires a certain time to get from the entrance to the exit and during this time it is stored in the cable. The carrier thereby transfers an arbitrary number of volatile, different symbols. As a hybrid and transition

---

177. Cf. DEPATIS, Patent no. US 2,263,902, p. 2: "The liquid container may be tubular in shape or narrow in one dimension and wide in another or wide in both dimensions."

178. Cf. Henry Fox Talbot, *The Pencil of Nature* (London, 1844).

179. Radio broadcasts publicly. Since no channel shields the transmission, anyone can read and write it, that is, tap it or disturb it. Interference and eavesdropping are two sides of the same coin as are the corresponding countermeasures, filtering and encryption.

180. The openness of the system proved to be especially problematic because any device that contained coils, like an electric motor, emitted radio waves. Household machines such as electric shavers, vacuum cleaners, kitchen machines, etc., became potential sources of disturbance.

between communication medium and storage medium the device connected two points; not two humans communicating with each other, but the reading and writing heads of the same circuit. The transformation of the signal into ultrasonic waves extended the time that the signal required to travel through the device. The idea of feeding it back into the system at the front probably originated in connection with the experience of acoustic echoes in telegraph and telephone systems.[181]

In communication, delay is a most unwelcome phenomenon, but from the angle described above, it can be considered as volatile, short-term storage. Long-term memory also originated from a new interpretation of a technical disturbance – feedback. Although extremely irritating when exchanging data through a channel that is supposed to be empty in order to send and receive information, feedback demonstrated the technical feasibility of storage in an ephemeral medium.

The tube containing liquid preserved a chain of pulses. Because the pulses travelled at the speed of sound, they were not only stored in space but in time as well. The distance from one crystal to the other and the time that the wave took to traverse this distance provided the basic beat. Furthermore, a clock subdivided this duration into smaller intervals so that symbols could be positioned within the flow of time: "This clocking is very important as it must keep the pulses in step as well as prevent degeneration of the pulses over a number of cycles."[182] It is not the pulses themselves that are refreshed repeatedly but their coincidence with the external rhythm. The division of the length of the tube determines the meaning of the square signals in the weak sense that they get an ordinal position, an address. Their location within the chain is then defined and can mean, for example, a power of 2. This achieves what Hegel regarded as the origin of numbers:

> "The *first* production of the number is the aggregating of the many [...] each of which is then posited as only a one – numbering. Since the ones are mutually external their representation is illustrated sensuously, and the operation by which number is generated is a process of counting on the fingers, dots, and so on. What four, five, etc., *is*, can only be *pointed out*."[183]

---

181. Alan Blumlein, the technical director of the EMI research group, began his career in 1924 as a telephone engineer at Bell Labs. There he developed a coiled cable that reduced mutual interference between channels (cross-talk) in long-distance telephone systems. In the early 1970s, the well-known telephone hacker John Draper, a.k.a. Captain Crunch, also sought the experience of continuity in volatile communication systems: "The hack, in this instance, refers to such technological stunts as having two phones on the table; talking into one, and hearing your voice in the other after a time-delay in which the original call has first been routed around the world." (Paul Taylor, *Hackers. Crime and the Digital Sublime* (London, 1999), p. 15). The acoustic coupling of the telephones' two receivers would have produced a storage medium.

182. T. Kite Sharpless, Mercury delay lines as a memory unit, in: *Proceedings of a Symposium on Large-Scale Calculating Machinery, 7–10 Jan. 1947*, ed. William Aspray (Cambridge, MA, 1985), pp. 103–109, p. 106.

183. Hegel, *Logic*, p. 206, translation slightly modified. And over a century later the Manchester Mark I actually did represent the data as dots. This could be an indication of the clairvoyant potential of systematic thought.

The Automatic Calculating Engine (ACE) that Turing projected in 1945 was based entirely on mercury delay lines; due to the war and administrative hurdles the machine only went into operation at the end of 1951.[184] Its programmers achieved "optimum coding" if they always read out packets of data from the tube at exactly the right moment and sent them to another tube thus avoiding waiting periods. What it actually means to develop software more in time than in space is described vividly and lucidly in Martin Campbell-Kelly's detailed article.[185] Moreover, the engineers had to control the ambient temperature closely as any variation affected the properties of the tubes.

Selecting and Indicating What Is Moving

In the transition from television to radar, an additional technique of mechanical memory was developed. After the carnage of World War I, in Western democracies there was decreasing acceptance of the high numbers of casualties on the battlefield, so military strategists in World War II sought to inflict intense, long-range strikes on the enemy civil population and infrastructure from the air and water.[186] The calculation of the trajectories of missiles was already aimed at determining the future position of an enemy target. Moreover, successful defence relied on early detection of the aggressor or "foreseeing" the deployment by some other means. Since the London Blitz and the bombardment of other British cities by German zeppelins and planes during the First World War, the British in particular had a vital interest in this.

After only a partially successful attempt in concentrating the noise of targets using massive concave "sound mirrors" made of concrete in order to localise distant targets, the British Government commissioned the NPL to investigate the possibility of using "death rays" to destroy enemy objects."[187] In his final report in February 1935, NPL's director Robert Watson-Watt summarised the investigations of his colleague Arnold Wilkins: "Although it was impossible to destroy aircraft by means of radio waves, it should be possible to detect them

---

184. A.M. Turing, Proposal for development in the Mathematics Division of an Automatic Computing Engine (ACE). Report to the Executive Committee of the National Physics Laboratory [1945], in: *The Collected Works of A.M. Turing. Mechanical Intelligence*, ed. Darrel C. Ince (Amsterdam, 1992), pp. 1–86. http://www.alanturing.net/turing_archive/archive/p/p01/P01-001.html.

185. M. Campbell-Kelly, Programming the Pilot ACE. Early programming activity at the National Physical Laboratory. *Annals of the History of Computing* 3 (1981): 133–162. See p. 150: "Unfortunately, optimum coding was a rather compulsive activity, and it was not always easy to have the self-discipline to stop at a point before the expenditure of programmer's time exceeded the saving in machine time."

186. Cf. Giulio Douhet, *The Command of the Air* [1921] (Washington, DC, 1983).

187. This idea, which appears in H.G. Wells *The War of the Worlds* of 1898, was revived by 78-year old Nikola Tesla in 1934. In an interview with the New York Times he claimed to be able to produce such rays and proposed protecting North America with an "invisible Chinese Wall of Defense" for only 2 million dollars. In a letter to the financier J.P. Morgan Jr., Tesla wrote: "One of the most pressing problems seems to be the protection of London and I am writing to some influential friends in England hoping that my plan will be adopted without delay." Cf. Margaret Cheney and Robert Uth, *Tesla. Master of Lightning* (New York, 1999), p. 144ff.

by radio energy bouncing back from the aircraft's body."[188] Two weeks later Wilkins gave a successful demonstration of the system to members of the Air Ministry. By the end of 1935 the tracking technology, which employed "echoes," i.e., back-scattered pulses, had a range of over 120 kilometres. In 1936 the government decided to protect the entire east coast of England and Scotland with a huge chain of radar towers, which was named "Chain Home."

Fig. 7: Radar towers on the east coast of Britain.[189]

In August 1937 the first British plane was equipped with a device to locate ships (RDF-2). Because Chain Home could not locate any low-flying objects with its long wavelength (in the meter range), by 1939 NPL collaborated with Alan Blumlein and his EMI laboratory team to develop the radar system GL ("gun-laying"), which used centimetric rays. Thus the hole in Chain Home was closed with "Chain Home Low." Three receiving antennas enabled manual determination of altitude, speed, and direction of targets. The apparatus was not only equipped with a CRT, which displayed the distance on the X-axis and the strength of the signal on the Y-axis (a so-called "A-Scope"), as of June 1940 it also had a PPI (plan position indicator), common today, which displayed objects present inside a given radius from a bird's-eye view. The intensity of the echo received by the rotating antenna determined the brightness of the light dots on the display, which were plotted radially from the middle of the screen outwards.

---

188. This phenomenon was actually discovered in 1904 by Christian Hülsmeyer; he patented his device as "Telemobiloskop." Cf. DEPATIS, Patent no. DE 165,546, and in general Robert Charles Alexander, *The Inventor of Stereo. The Life and Works of Alan Dower Blumlein* (Oxford, 2000), p. 229ff. In my depiction of the early development of British radar I follow Alexander's work.

189. Richard Townshend Bickers, *The Battle of Britain. The Greatest Battle in the History of Air Warfare* (London, 1990), p. 87.

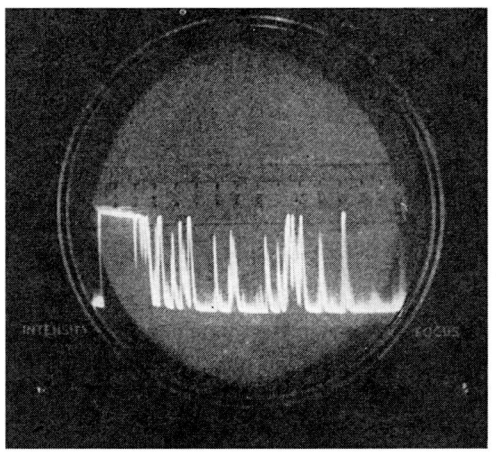 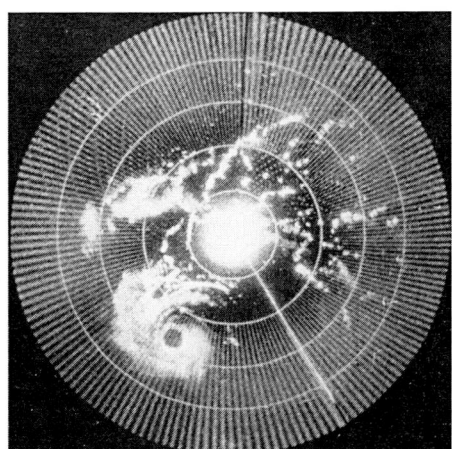

Fig. 8: Display of data on the A-Scope and PPI.[190]

In connection with the above, the newly formed Radar Research Group of the Royal Air Force (later re-named Telecommunications Research Establishment – TRE) developed a technology to distinguish between friendly and enemy planes, which was based on a transponder and was called Identification Friend or Foe (IFF). In 1939 the young engineer Frederic Williams perfected the device and the electrical firm Ferranti Ltd. in Manchester produced it.[191] Williams also played a prominent part in improving mobile radar for aircraft, which tracked and intercepted other objects in the air, Airborne Interception (AI). When the technicians realised that fields, cities, and other regions equally reflected waves with individual characteristics, from the end of 1942 TRE and EMI began work on a target recognition system called H2S. Until its completion in January 1943, 90% of bombers missed their targets. After the destructive attacks on Hamburg, Leipzig, and Berlin, in early 1944 Adolf Hitler admitted, "with regard to technical inventions in 1943 the balance may have tipped in favour of our enemies."[192]

---

190. Louis N. Ridenour, ed., *Radar System Engineering. Massachusetts Institute of Technology Radiation Laboratory Series, Vol. 1* (New York, 1947), p. 165; Wikipedia, "Radar."

191. Alexander, *Inventor of Stereo*, p. 256ff. Williams and Blumlein met in the autumn of 1940; cf. p. 277: "Blumlein made a great impression upon Williams, and the latter was said to have never lost his admiration for him. Williams was particularly moved by Blumlein's approach to engineering and circuitry at EMI, and recognised with greater clarity than he had ever done before that with the right approach circuits could be designed. [...] Following this meeting with Blumlein, Williams' approach was quite changed and he too adopted designability as the driving force behind his work." To understand this rather incomprehensible remark some sixty years later, one has to remember that at that time it was by no means clear how precisely the properties of electronic components, which were only just being developed, could be shaped and thus also the circuitry. See also the concluding paragraphs below.

192. Ulrich Kern, *Die Entstehung des Radarverfahren. Zur Geschichte der Radartechnik bis 1945* (Ph.D. diss., University of Stuttgart, 1984), p. 245ff. See also Bernard Lovell, *Echoes of War. The Story of H2S Radar* (Bristol, 1991).

In June 1943, three members of the team, including the 39-year-old Blumlein, were killed in a plane crash while on a test flight.

Increasing improvement of the sensing power of radar brought problems with it, which were discussed under the headings of "permanent echoes" and "ground clutter." Stationary objects like mountains and buildings reflected the pulses, concealed moving objects, and irritated the operator by generating a lot of irrelevant information.[193] Just as with television, it was necessary to single out, or filter what was of interest out of the uniformly recorded data: "Often a radar system sees too much, rather than too little."[194] Beginning in 1940, William S. Elliott at the Air Defence Research and Development Establishment (ADRDE) worked on adapting William Percival's system to long-wave radar. A delay line stored the echo in order to subtract it from the next one received.[195] Paradoxically, it was the desire to extract moving objects from the data that led to the necessity of storing the patterns. Only when the two dimensions of the screen were extended by the third dimension of time was it possible to subtract the past signals from the present ones, and in this way to filter out what was constant and thus undesirable, i.e., that which did not cease to write itself continually and identically.

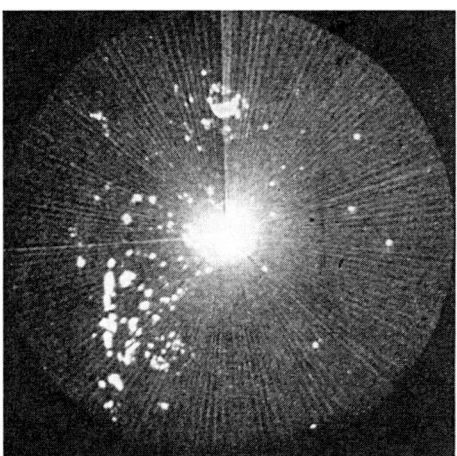 

Fig. 9: PPI without (left) and with Moving Target Indication (right).[196]

193. Cf. Figure 9.

194. Ridenour, *Radar System Engineering*, p. 124.

195. See NCUACS, Guide to the Manuscript Papers of British Scientists. Elliott, William Sydney (1917–2000), Computer Engineer, CSAC no. 121/7/03 (Bath, UK, 2003; online: http://discovery.nationalarchives. gov.uk/details/rd/N13650598): "His Ph.D. studies at the Cavendish Laboratory, Cambridge, were interrupted when he joined the wartime Air Defence Research and Development Establishment at Christchurch, Hampshire, later moving to Malvern, Worcestershire. During this period he worked on radar systems, developing an interest in pulse-type electronic techniques. Projects included the use of delay lines to cancel out interference of stationary 'clutter' in radar signals, to distinguish a moving target."

196. Ridenour, *Radar System Engineering*, p. 627.

In 1942 Britton Chance from the Radiation Laboratory of the Massachusetts Institute of Technology (MIT) in Cambridge, USA, visited the British engineer Frederic Williams, who was now established at TRE, to initiate an exchange concerning progress in radar research between the two countries: "I was to learn everything they were doing, and to tell them everything I was doing."[197] In their report of 1944, which was classified information until 1960, the American engineers Robert A. McConnell, Alfred G. Emslie, and F. Cunningham, who worked on Moving Target Indication on the American side, came to the following conclusion about the British work:

> "The British have used a water delay line in long-wave radar. Its success has been limited by bandwidth, attenuation, and temperature problems. [...] It is a characteristic limitation of the delay line that the system pulse rate must be precisely constant. [...] To avoid this limitation, a static storage method is needed – one which will preserve the video pattern for an indefinite period, ready for comparison with the succeeding video pattern. The television mosaic provides a means by which this may be accomplished."[198]

The authors also pointed out that the simple subtraction procedure led to "blind regions of no response whatsoever" in the radar image. Like Percival's apparatus, the technique switched off part of the image so that moving objects could no longer be perceived there. What was required, they said, was "selective elimination of ground echoes and the maintenance of a maximum sensitivity to moving targets at the same radar range."[199] The engineers succeeded in doing this by analysis of the exact wave form of echoes of successive pulses. They displayed and temporarily stored these on a CRT whose signal plate was connected to a video amplifier.[200] Through the Doppler effect, the polarity and amplitude of the echoes of moving objects changed in continuous wave radar from pulse to pulse and the echoes of static objects remained the same. The technicians measured the change as a positive charge at the moment the segment of the corresponding wave was displayed, then amplified it and marked the moving targets with a light

197. Andrew Goldstein, Britton Chance, Electrical Engineer, an Oral History (New Brunswick, NJ, 1991; online: http://ethw.org/Oral-History:Britton_Chance).

198. Robert A. McConnell, Alfred G. Emslie, and F. Cunningham: *A Moving Target Selector Using Deflection Modulation on a Storage Mosaic. M.I.T. Radiation Laboratory Report No. 562* (Cambridge, MA, June 1944), p. 4. For a detailed understanding of the television set, McConnell et al. refer the reader to *Television* by Zworykin and Morton: Vladimir K. Zworykin and George A. Morton, *Television: the Electronics of Image Transmission* (New York, 1940). The pioneer of television in North America, Vladimir Zworykin, also studied under Rosing; see fn. 175.

199. McConnell, *Moving Target Selector*, p. 1.

200. Storing data on the radar screen merely made a process explicit that had always been present here: because it was only partially possible to shield the receiving antenna against the emitted radiance, in PPI representation the centre was always lit and not sensitive to other objects (cf. Figure 9). The "I=I," the "representation which must be capable of accompanying all other representations, and which in all consciousness is one and the same" manifested itself technically. Quotation: Immanuel Kant, *Critique of Pure Reason*, trans. Norman Kemp Smith (London, 1929), p. 153. As in the delay line, feedback produced the continuity of a signal.

dot. In addition, the MIT Radiation Laboratory worked on systems that used the intensities of a two-dimensional area for display and calculation.[201] Instead of the direct signal on the monitor, which before had displayed the actual radar echoes received, the indicator generated arbitrary symbols.

Signals of Angels and Symbols of Nothing

The rays received from the external world allowed for the precise algebraic processing of successive waveforms, which visualised objects that had previously been imperceptible to the naked eye. This was achieved by eradicating the traces of objects from the display that were also real but unimportant because they were constant. In this process, the engineers used a signal generator for the information that they wished to pull out and display on the CRT. Photography and television were touted as technologies that faithfully recorded reality. Radar, however, broke the apparent unity of reality and its representation apart, because it programmatically manipulated the image. The pictures were not a faithful record of the rays received; they merely represented the initial data for filtering, that is, the algebraic calculation of the image. Slowly but surely, algorithms were beginning to determine what was considered as real.

According to Baudrillard, who provides a modern paraphrase of Hegel's dialectics of the essence and its appearance, the relationship between reality and its representation develops in four stages: "1. It [the representation] is the reflection of a basic reality. 2. It masks and perverts a basic reality. 3. It masks the *absence* of a basic reality. 4. It bears no relation to any reality whatever: it is its own pure simulacrum."[202] As of 1941, the combination of highly sensitive sensors and imaging produced by calculations resulted in the appearance of clear-air echoes (so-called "angels") on Doppler radar screens that the pilots flying past were unable to identify the source of. Naturally, this astonished and baffled the operators. These signals hallucinated by the technical system, which correspond to Baudrillard's third, "magic" phase, fanned the flames of discussions about unidentified flying objects and extraterrestrial alien life forms in the 1950s.[203]

The director of the Radiation Laboratory, Louis Ridenour, suggested as early as 1944 that the entire body of knowledge on radar accumulated in operations

---

201. McConnell, *Moving Target Selector*, p. 6ff.

202. Jean Baudrillard, *Selected Writings*, ed. Mark Poster (Cambridge, UK, 1988), p. 170.

203. See James E. McDonald, Meteorological factors in unidentified radar returns, in: *14th Radar Meteorology Conference, November 17–20, 1970* (Boston, 1971), pp. 456–463, p. 456: "Similarly, productive research on what ultimately proved to be a wide variety of types of 'radar angels' stemmed from efforts to account for peculiar echoes not identifiable as aircraft or precipitation or ground returns." The observation of radar angels later proved productive in radar meteorology. Interestingly, the engineer Robert A. McConnell mentioned above began to be interested in parapsychological phenomena while he was still at the Radiation Lab. He conducted first experiments in 1947 and from then onward, concentrated exclusively on this research. In 1957, the US Parapsychological Association elected him as their first president. See the Association's website at http://archived.parapsych.org/members/r_a_mcconnell.html.

research during the war be gathered together in one large work. He probably wanted to convey the impression that the main developments had taken place in the USA. Britton Chance contacted his British colleague Frederic Williams and invited him to work on two of the volumes in the now famous series, which ultimately numbered 28. To this end, the British engineer visited the Radiation Laboratory in 1945 and 1946, and there he learned about the experiments of McConnell and his co-workers in which they stored radar data on CRTs. The device did not achieve the robustness necessary for application in the field and was evidently abandoned. In their contribution to the book series of 1947, Emslie and McConnell only mention their own research briefly with a single sentence: "It is also possible to delay the signal by means of a 'storage tube.' [...] The supersonic delay line was used as a delay device in the MTI systems that have had the most thorough testing; its use is therefore assumed in what follows."[204] The Americans had overlooked the decisive fact that by using the time gained by short-term storage for refreshing the data just read, memory could be extended indefinitely: "Looking back, it is amazing how long it took to realise the fact that if one can read a record once, then that is entirely sufficient for storage, provided that what is read can be immediately rewritten in its original position."[205]

With the USA's nuclear attacks on Hiroshima and Nagasaki in August 1945, World War II came to an abrupt end, and thus also the intensive research on radar:

'My interest in computers [...] was directly caused by the atom bomb. Substantially overnight this event converted a mass of radar experts with endless problems for which they were seeking solutions, into a mass of experts with endless solutions and no problems, for in those days we were naive enough to believe that the end of war meant the beginning of peace."[206]

Fig. 10: Williams' CRT Store being used slightly improperly as a visual medium.[207]

In December 1946 at TRE, a few months after his return from the USA, Williams successfully stored a single bit stably on a CRT.[208] At the University of Manchester, with the help of his assistant Tom Kilburn, he improved the apparatus, and in 1948 was able to represent up to 2048 "digits" on a screen. Hallucinatory signs, which only indicated angels, because there were no more enemies

204. Ridenour, *Radar System Engineering*, p. 631.

205. F.C. Williams and T. Kilburn, The University of Manchester computing machine, in: *Faster than Thought*, ed. Bowden, pp. 117–129, p. 117.

206. Williams, Early computers, p. 327. All these things – and contrary to Heraclitus – were spawned by the end of the war and of the pressure of immediate applicability.

207. Williams and Kilburn, A storage system, p. 184.

208. T. Kilburn, From cathode ray tube to Ferranti Mark I. *Computer Resurrection. The Bulletin of the Computer Conservation Society* 1 (1990): 16–20, p. 16.

in the skies, thus changed into symbols of nothing; pure signs that could take on any arbitrary meaning.

A "pick-up plate" which caught the electrons covered the front side of the screen. As in the delay line, the writing head of a circuit replaced the human observer. Depending on the current state of the data charges (on or off), the bombardment of dots immediately in the vicinity resulted in signals of different polarities on the plate, which – as in McConnell's construction – were amplified and used to re-write the information that had just been read and to switch the echo to endless. The proximity of the bombarded "pixels" produced interference and polarity changes similar to those in continuous wave radar when moving targets were observed. In Williams' own words – marked by the experience of war – this resulted in the following properties of the artefact:

> "(a) Either of two states of charge may be left at will at a given spot on the c.r.t. face. These states are (i) a well of full depth, by bombarding the storage spot, ceasing the bombardment and not bombarding any other spot in the vicinity, or (ii) a partially filled well, by bombarding first a storage spot, and then another spot in the vicinity before ceasing bombardment. (b) Charge distributions will be maintained for a time – a few tenths of a second – depending on surface leakage. (c) Renewed bombardment […] of the storage spot will give, at the instant of recommencing bombardment, a negative signal from the amplifier in case (a) (i), or a positive signal in case (a) (ii). (d) Bombardment of spots displaced by more than 1.33 spot diameters from a given spot has no influence on the potential distribution at that spot."[209]

As in battle, waves of bombardment on locations and their environs produce memories because they leave craters or "wells" behind them. What followed was the construction of the first computer, the Manchester "Baby Machine," which in June 1948 ran a programme to calculate the highest factor of 2^{18}. It was built for the simplest of reasons: "[T]he only way to test whether the cathode ray tube system would work with a computer was, in fact, to build a computer."[210] From 1949 to 1950, the computer was extended and modified on a daily basis, without a master-plan, to become the Manchester Mark I. It was replaced in 1951 by the Ferranti Mark I, which like Williams' radar equipment was built by the Manchester firm of the same name.[211]

---

209. Williams and Kilburn, A storage system, p. 188.

210. Kilburn, From cathode ray tube, p. 18.

211. F.C. Williams, T. Kilburn, and Geoffrey C. Tootill, Universal high-speed digital computers. A small-scale experimental machine. *Proceedings of the Institution of Electrical Engineers, Part II* 98 (1951): 107–120. Cf. Campbell-Kelly, Mark I, p. 130f.

Fig. 11: The Manchester University Mark I, 1949.[212]

"The fact that his [Turing's] Universal Machine had materialised mathematics allowed the reverse, to mathematise matter."[213] The inverse of the sentence is also true: the condition for representing changeable symbols lay in the fact that the scientists no longer understood natural phenomena, such as electricity, as fate and fact to be grasped descriptively, but as material that could be formed in any number of ways and in which they could write chains of simple symbols. In the first of the volumes on radar on which Williams collaborated and which bears the telling and modest title "Waveforms," he said goodbye to the traditional way of looking at waves:

> "Previous treatment of waveforms has been directed mainly to sinusoids and the various manipulations that can be performed on them. […] In approaching the subject matter of this book it is preferable to make a clean break with the traditional approach. […] The waveforms that will be considered are not sine waves, but square waves, pulses, and even more complicated shapes."[214]

Originally used and understood as energy to power light bulbs or drive machines, and later as an analogue transmission medium, an interpretation and technology of electricity emerged which made it possible to form waves freely in different shapes and to represent symbols in them which transformed, even redundant echoes of love.

It is a good question as to why one of the very early programmes on the first computer generated letters of this kind, that is, love-letters. According to Freud, love is a phenomenon that more than any other is characterised by projections, and more than any other the love-letter is a genre that invites one to suppose the feelings and thoughts that lie behind it. Goethe, for example, once made

---

212. See Lavington, *Early British Computers*, p. 38: "The Mark I was built out of war-surplus components with an enthusiasm that left little time for tidiness!"

213. Friedrich Kittler, *Unsterbliche. Nachrufe, Erinnerungen, Geistergespräche* (Munich, 2004), p. 78.

214. Britton Chance, Vernon Hughes, Edward F. MacNichol, David Sayre, and F.C. Williams, eds., *Waveforms. Massachusetts Institute of Technology Radiation Laboratory Series, Vol. 19* (New York, 1949), p. 8.

the cynical suggestion that love-letters should be formulated in a completely cryptic way, so that the recipient could project whatever she liked into the text.[215] With the transformation of signals into signs of nothing, however, precisely this operation is necessary. Meaning can only be given to the "mad dance" of the storage spots on the Mark I and all the computers that came afterwards from outside.[216] Without the projection that endows meaning, the computer itself merely separates and unites, writes and deletes – dots.

215. Cf. Johann Wolfgang Goethe, *Briefwechsel mit Marianne und Johann Jakob Willemer*, ed. Hans-J. Weitz (Frankfurt a. M., 1986), p. 26f.: "One would do best to write something completely incomprehensible so that friends and lovers would have complete liberty to put true meaning into it." I am indebted to Wolfgang Pircher for this information. Translation mine, D.L.

216. Williams, Early computers, p. 330.

Enigma Rebus
Prolegomena to an Archaeology of Algorithmic Artefacts

The Consolation of the House Father

"At first glance it looks like a flat star-shaped spool for thread, and indeed it does seem to have thread wound upon it; to be sure, they are only old, broken-off bits of thread, knotted and tangled together, of the most varied sorts and colours. But it is not only a spool, for a small wooden crossbar sticks out of the middle of the star, and another small rod is joined to that at a right angle. By means of this latter rod on one side and one of the points of the star on the other, the whole thing can stand upright as if on two legs."[217]

With these words, the Czech writer Franz Kafka (1883–1924) describes a mysterious being named Odradek in his short story *The House Father's Concern*, written around 1917 and first published in 1919. The strange apparatus moves around autonomously, talks like a child, and occasionally laughs dryly. If this were a report about a robot that operates on a familiar terrain and responds "Odradek," "No fixed abode" and "Haha" if spoken to, the text would not be very surprising. A similar anthropoid, "Elektro," was built twenty years later by the Westinghouse Electric Corporation and exhibited at the 1939 World's Fair in New York. It could walk by voice command, talk (it incorporated a 78-rpm record player), smoke cigarettes, blow up balloons, and move its head and arms. Similar statues with less functionality (or "features") were developed by Hero of Alexandria in Ancient Greece, by Al-Jazarí in Upper Mesopotamia at the

---

217. Franz Kafka, The cares of a family man [1919], in: *The Complete Stories* (New York, 1971), p. 428. The story "Die Sorge des Hausvaters" was first published in: *Selbstwehr. Unabhängige jüdische Wochenschrift* 13. 51/52 (19 December 1919, Chanukkah issue). "The House Father's Concern" is a more accurate translation of the title, because the German "Sorge" in its intransitive form rather means to worry and not to care, and "Hausvater" does not necessarily imply that the person is married. He could well be a kind of warden.

beginning of the thirteenth century, and subsequently by many others.[218] Odradek is a simpler artefact constructed from fewer components – two, but has in a mysterious way acquired biological properties (cf. Figure 1). Since antiquity mobility has been regarded as the *differentia specifica* of living beings, and the ability to talk and to laugh that of humans: "That man alone is affected by tickling is due firstly to the delicacy of his skin, and secondly to his being the only animal that laughs."[219]

The main property that unites Odradek with the creature possessing language is that he does not serve any purpose, as a tool or instrument. "Now I say that the human being and in general every rational being *exists* as an end in itself, *not merely as a means* to be used by this or that will at its discretion."[220] This state of purposelessness, which was historically often interpreted as freedom, can take different forms. Estragon and Vladimir, the main characters in Samuel Beckett's *Waiting for Godot* (1952), deduce from the fact that they exist, but have nothing to do, that they are waiting, and from waiting that somebody will arrive. This interpretation of the Irish writer's play is offered in the lucid anti-Heideggerian essay *Being without Time* by the Austro–German philosopher Günther Anders (a.k.a. Stern).[221] Similarly, it could be conjectured that the little device Odradek has concluded, qua Kant's definition, that because it does not fulfil a function it is a human being and should henceforth walk about, communicate, and laugh occasionally.

According to Anders, one characteristic of the parable, the genre to which the small text of Kafka very likely belongs, is poetic inversion. To express that humans sometimes show animal behaviour, it creates the alienation effect ("Verfremdung," in German) of portraying animals as humans by exchanging subject and predicate of the proposition.[222] In the case of Odradek this linguistic technique would result in a statement surpassing Julien de La Mettrie's

218. Hero Alexandrinus, *The Pneumatics of Hero of Alexandria* [ca. 62 AD], trans. Marie Boas Hall (London, 1971); Ibn ar-Razzáz Al-Jazarí, *The Book of Knowledge of Ingenious Mechanical Devices* [ca. 1206], trans. Donald R. Hill (Dordrecht, 1974); cf. Chris Hables Gray, Steven Mentor, and Heidi J. Figueroa-Sarriera, eds., *The Cyborg Handbook* (London, 1995), p. 89.

219. Aristotle, *On the Parts of Animals* [ca. 350 BC], trans. William Ogle (Oxford, 1882), p. 84.

220. Immanuel Kant, *Groundwork of the Metaphysics of Morals* [1785], trans. Mary Gregor (Cambridge, 1998), p. 37. Also in Kafka's "Penal Colony" a machine plays a central role, and the transformation of an object into some kind of human being, appearing also in "Blumfeld, an Elderly Bachelor," is but one of several metamorphoses transgressing ontological boundaries in his oeuvre. In "The Metamorphosis," a family's son turns into an insect; in "A Report to an Academy" an ape becomes a cultivated subject, etc. The classical model for these is obviously Ovid's *Metamorphoses*.

221. Günther Anders, Sein ohne Zeit. Bemerkungen zu Samuel Becketts Stück "En attendant Godot." *Neue Schweizer Rundschau* 21 (1953/1954): 526–540. Included in: G. Anders, *Die Antiquiertheit des Menschen. Band 1: Über die Seele im Zeitalter der zweiten industriellen Revolution* (Munich, 1956), pp. 213–231. English translation in: Martin Esslin, ed., *Samuel Beckett. A Collection of Critical Essays* (Englewood Cliffs, NJ, 1965), pp. 140–151. Anders studied under Martin Heidegger and was married to Hannah Arendt from 1929 to 1937, after his teacher's love affair with her in 1925/1926.

222. More on inversion, especially in Hegel and Freud, can be found in D. Link, *Poesiemaschinen / Maschinenpoesie. Zur Frühgeschichte computerisierter Texterzeugung und generativer Systeme* (Munich, 2007; in German), pp. 66–69.

infamous *L'homme machine*: Human beings are not only machines, but even worse, senseless and absurd apparatuses, because they cannot be used to any end.[223] The specific difference of the *zõon* would no longer be *lógon échon*, but the defect to be unable to serve any purpose. Anders derives an analogue conclusion in his book on Kafka:

> "If the human being seems 'inhuman' to us today, it is not because he possessed an 'animal' nature, but because he is pushed back into *object functions*. Therefore, the contemporary poet has to invent fables in which objects appear as living beings to denounce the scandal 'humans are objects.' And Kafka has drawn this consequence."[224]

Similar to Beckett's *clochards*, Odradek does not inhabit a house he could be the father of ("No fixed abode"), serves no particular purpose, and pursues no goal. However unlike them, since he does not wait for the sudden advent of transcendence, he eternally wanders around senselessly like a will-o'-the-wisp, and returns like a ghost.

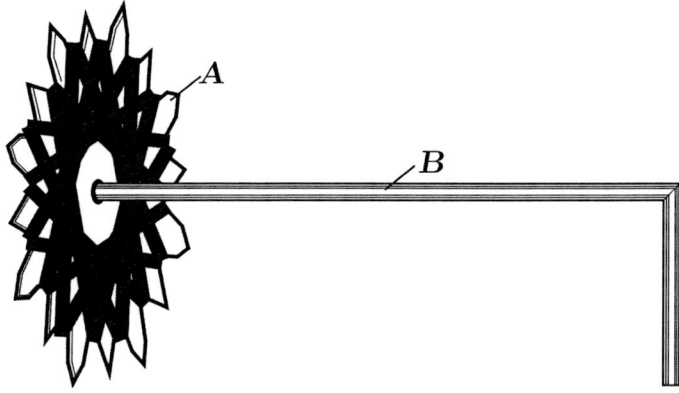

Fig. 1: Reconstruction of Odradek from descriptive text; A: star-shaped spool, B: wooden crossbar.

By contrast the father guarding the house fulfils the minimal function of being there, keeping his eyes open, and taking care of problems if necessary. That he is located at the border between duty and futility endows him with the time to reflect on what he encounters, an autonomous artefact with no purpose at all. He does not care about Odradek, and is only concerned about himself. He suspects that even if the senseless thing copies several characteristics of human beings,

---

223. Julien Offray de La Mettrie, *L'Homme Machine* (Leyden, 1748). The mechanistic view of human beings seems reasonable, because the basis of their complex behaviour, including thought, is obviously material.

224. G. Anders, *Kafka pro und contra* (Munich, 1951), p. 12f. (Translation mine, D.L.) Throughout this essay based on a lecture held in 1934, Anders calls Odradek "Odvadek," without specifying any reason. In Hungarian, "odvas" means "dry-rotten, scooped out, decayed."

it might be immortal or at least more durable than he himself, because it does not wear itself out in the pursuit of an aim. One of Anders' main concepts, the "Promethean shame," bases on a similar train of thought. Human beings feel embarrassed vis-à-vis their products, because they seem perfect compared to themselves and possess a higher durability, which is brought about by "platonisation." Anders takes the light bulb as an example. If it burns out, he argues, it can be replaced by another one in a process of "industrial reincarnation," because it is only one out of millions of objects realising the same idea, which represents the essence of the product.[225]

Marshall McLuhan called "electric light [...] pure information" and "the medium without a message."[226] The bulb neither fits the category of "tool" or "instrument," nor that of "medium." It does not allow one to manipulate other objects, to produce scientifically interesting effects or to observe yet unknown aspects of nature, nor does it convey a message of any kind. Hence the more general term "artefact" will be used in this book to denote things that have been skilfully produced to serve a certain purpose, either directly or indirectly.[227] These objects possess a second, symbolic level or a "matrix," as Anders calls it, which is constituted by the concept of a desired function (artificial light in the case of the bulb) and the idea of an implementation, the association of material elements in a certain *technical form*.[228] Seen from a momentary perspective, this in fact results in an "eternalisation." But from a more historical perspective, its effects might serve as a consolation for the house father as well as for the human beings embarrassed by "Promethean shame": Different from a piece of wood, artefacts disintegrate on several levels and may suffer, as it were, several different deaths:

(1) The desired function might become obsolete, due to change of needs, fashion, or realisation of its impossibility. An example of the last class of aims is the making of gold from less valuable material undertaken by alchemists for several hundred years, which lead to the Western recipe for porcelain amongst other things.

(2) Another artefact might be invented to better implement the desired function, thus rendering the old one obsolete, as can be seen in the sequence of inventions for artificial illumination: candle – kerosene lamp – gas lamp – incandescent light bulb.

(3) The material basis might disintegrate to a point where it becomes unusable: parts might fall off or wear out, etc.

---

225. Anders, *Antiquiertheit*, p. 51.

226. Marshall McLuhan, *Understanding Media. The Extensions of Man* (New York, 1964), p. 8.

227. Latin *ars* – skill and *facere* – to make: something skilfully produced.

228. Cf. the etymological elaborations of Heidegger on the connection of "Zeug" (artefact), "Zeigen" (pointing) and "Zeichen" (sign), in Martin Heidegger, *Being and Time* [1927], trans. John Macquarrie and Edward Robinson (New York, 1962), pp. H 68–83.

The Idealisation of Artefacts

The symbolic level of the artefact is realised by configuring a certain material association of individual components. The statement of McLuhan's media theory that "the 'content' of any medium is always another medium" should be reformulated in this context: The components of an invention are in most cases artefacts devised earlier.[229] The light bulb owes its existence to the craft of glassblowing, the invention of the mercury vacuum pump, the generation and transmission of electricity, etc. If the association reaches a certain complexity, for example, in mechanical apparatus, an uninitiated observer can no longer easily gauge the intended function from the form, especially if it serves an uncommon, specialised purpose. Consequently, Anders diagnoses human beings to be alienated from their artefacts:

> "The object ['Odvadek'] reminds us of all sorts of items and machines which modern man has to handle day in, day out, even though their performance seems to have nothing to do directly with the *needs* of human beings. Thousand-fold, the contemporary human being is confronted with apparatuses whose purpose is unknown to him, and to which he can only maintain 'alienated' relationships, because their relations to the human system of needs are indefinitely mediated."[230]

Incandescent light bulbs stand at the beginning of a development in which this quality of being enigmatic becomes aggravated. The German economist and philosopher Alfred Sohn-Rethel called them "inscrutable spiritual beings" in his brilliant essay on the Italian way to deal with technology, and noted that in this southern country, these artefacts "which always leave one wondering if they are even of this world [...] flow together uninhibitedly with the nimbus of the religious powers, and the festive Osram bulb is united, in Neapolitan saintly images, with the Madonna's aureole."[231] With mastery of the invisible force of electricity, artefacts started to get "platonised," or better yet, idealised in a sense different from Anders'. The light bulb no longer achieved a material effect in the world, but merely a visual one, and one component of the implementation was invisible – the electric current. The "medium without a message" developed further into media with a message with the advent of imaginary artefacts such as film projectors, when semitransparent material was put in front of it. The success of these apparatuses and the market for immaterial commodities they made possible at first highly astonished traders of ordinary goods. To them, these machines, which produced "nothing," must

---

229. McLuhan, *Understanding Media*, p. 8. This point is well illustrated by the technical trees incorporated in computer strategy games like "Civilization" by Sid Meyer (1991), even if sometimes, due to misconception, it is possible to invent the tank before the wheel.

230. Anders, *Kafka*, p. 11.

231. Alfred Sohn-Rethel, Das Ideal des Kaputten. Über neapolitanische Technik [1926], in: *Das Ideal des Kaputten* (Bremen, 1990), pp. 33–38, p. 35f. English translation online, The Ideal of the Broken-Down, http://www.formundzweck.com/eng/autoren.php?S+Sohn-Rethel+Alfred. Thanks to Alexandre Métraux for bringing this article to my attention.

have seemed as absurd as Odradek. Before happening upon a cinetoscope show, the cloth merchant Carl Laemmle "had never seen such a rush of customers, and started to count the masses, marvelling," thereafter he decided to take part in the new business. His Independent Moving Picture Company merged with other studios in 1912 to become the Universal Film Manufacturing Company, one of the biggest film production facilities in the early years of the medium.[232] Siegfried Zielinski has pointed out that already at the beginning of the eighteenth century, both stock dealers and *laterna magica* projectionists were ridiculed as "traders of wind."[233]

In their implemented form, artefacts possess an ambiguity that allows them to be reinterpreted experimentally. In 1904, the British electrical engineer and physicist Ambrose Fleming slightly modified the material association of the light bulb by incorporating a second wire into the vacuum tube and created the first component that was no longer electric, but electronic (cf. Figure 2). Fleming had worked for Edison's company in London, where he became interested in the "molecular shadow" inside lamps, already noted by the American inventor, which was caused by the emission of negative particles from the incandescent filament.[234] In his "Instrument for converting alternating electric currents into continuous currents," electricity only flowed in one direction, from the heated cathode to the anode, in the form of a "cathode ray." Its original function was to make "feeble electric oscillations [of alternating current], such as are employed in [...] telegraphy" "detectable by and measurable with ordinary direct current instruments."[235] The rectifier or diode was incorporated into the Marconi-Fleming valve receiver for wireless telegraphy from 1905 onward.

Two years later, the US American inventor Lee de Forest added another electrode to the setup and created a second artefact with an invisible function, the "Audion."[236] The electron cloud at the heated cathode was attracted to the anode when a feeble positive charge was applied to a control grid located between the two. In this way, currents of various strengths could be switched without any material, or inert components interfering. Since the flow to the anode was proportional to the charge on the grid, the device could in principle also be

---

232. René Fülöp-Miller, *Die Phantasiemaschine. Eine Saga der Gewinnsucht* (Berlin, 1931), p. 11f. The next such astonishment might occur when the "industry of consciousness" resulting from this trade of immaterial goods realises that digitisation and instant electronic communication have rendered its products unsaleable. Since they can be copied and communicated at will, virtual objects can only be transacted once between the producer and the market as a pool of common intellectual property. Because of the lack of precedents, it cannot be estimated whether they still constitute commodities.

233. Cf. Siegfried Zielinski and Silvia Wagnermaier, eds., *Variantology 1. On Deep Time Relations of Arts, Sciences and Technologies* (Cologne, 2005), p. 144. The author is also indebted to S. Zielinski for suggesting the form of prolegomena for the present chapter.

234. Cf. Thomas A. Edison, Electric lamp (Patent no. US 223,898, 4 November 1879) and Gerald F.J. Tyne, *Saga of the Vacuum Tube* (Indianapolis, 1977), pp. 31–51.

235. John Ambrose Fleming, Instrument for converting alternating currents into continuous currents (Patent no. US 803,684, 19 April 1905, first applied for in England, GB 24,850, 16 November 1904), p. 1. All patents quoted are accompanied by their date of application.

236. Lee de Forest, Device for amplifying feeble electrical currents (Patent no. US 841,387, 25 October 1906) and Space telegraphy (Patent no. US 879,532, 29 January 1907).

used as an amplifier. It was built into a number of radiotelephone sets that his company sold to the US Navy in 1907.[237] Before copying and extending Fleming's artefact, de Forest had tried to achieve the same effect by modifying Bunsen burners. That he developed the Audion more through experiments than by concepts is further evinced by the fact that he did not understand the reason why his invention worked: "[T]he explanation is exceedingly complex and at best would be merely tentative."[238]

The light bulb created an immaterial, but visible effect, whereas in the electron tube it was immaterial and invisible, constituting an "ideal artefact," in which some parts – and more importantly its purpose – were located in the realm of sub-atomic particles as well as in the waves that could be formed out of their beams. In the space without air, the material resistance that had defined natural matter since antiquity was suppressed and, consequently, the component reacted almost without delay.

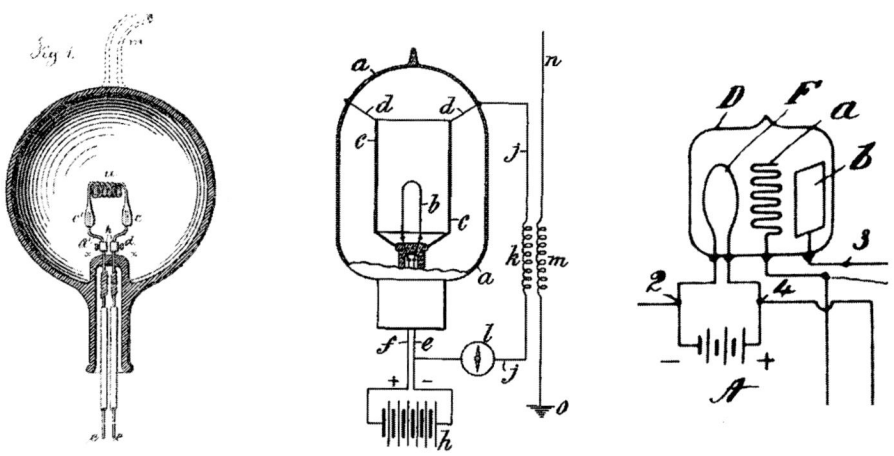

Fig. 2: Three examples of idealised artefacts: light bulb, 1879; diode, 1904; triode, 1907.

As a result, the implemented functions looked increasingly identical from the outside and could only be differentiated by an expert, using special measuring equipment which had evolved out of the same scientific and engineering movement, like Karl Ferdinand Braun's instrument for "the demonstration and study of the temporal sequence of variable currents" invented in 1897.[239] The component that invisibly performed its task itself became opaque

---

237. Tyne, *Saga*, pp. 52–72.

238. De Forest, Space telegraphy, p. 1.

239. Ferdinand Braun, Über ein Verfahren zur Demonstration und zum Studium des zeitlichen Verlaufes variabler Ströme. *Annalen der Physik und Chemie. Neue Folge* 60 (1897): 552–559. The predecessor of today's "oscilloscope" antedates the first "ideal artefact," a "diode" in modern terminology, by seven years.

and transformed into a black box. If an external observer was not instructed by the documentation or the cultural knowledge around him, it might have seemed to him as absurd as Odradek, because its significance could no longer be read from its form. The enigmatic quality of artefacts increased even more when these components were in turn integrated into bigger, more complex devices like radio receivers. They encapsulated functionality into single elements that served distinct purposes, but created the desired effect through their interplay. Similar to a rebus, where the meanings of pictures of objects and letters have to be combined in order to guess at the sentence represented, when confronted with complex idealised machinery, the observer first has to find out the sense of the different components, and then combine them to derive the overall purpose (cf. Figure 3).[240]

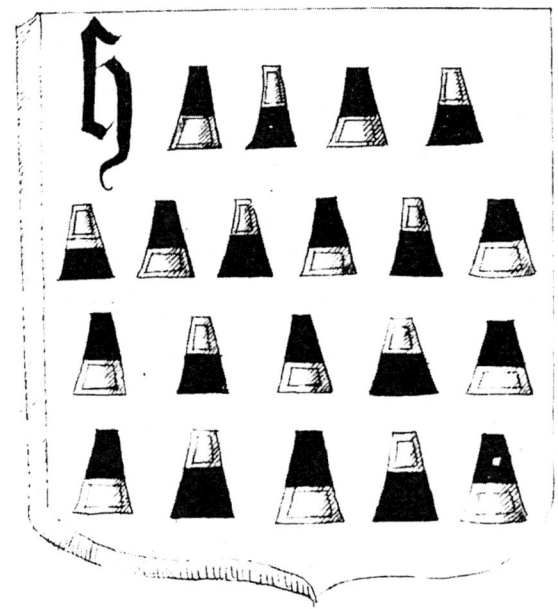

Fig. 3: A riddle from the Rébus de Picardie, 1506.

To enable their possessors to deduce and use the functionality of complex and, later, idealised artefacts, their constructors complemented them with manuals. For similar reasons, Great Britain required a written specification for the application of patents in the early eighteenth century, and the explanatory text on the "new and useful improvements" was soon illustrated by additional technical

---

240. Cf. Jean Céard, ed., *Rébus de Picardie: les Manuscrits f. Fr. 5658 et 1600 de la Bibliothèque Nationale* [~1499–1506] (Paris, 1986), no 15-XIV. "H" (letter name pronounced "ash" in French) is followed by "vingt pions," twenty pawns: "ch'vin pions" – "This wine we drink."

drawings.[241] Whereas for natural objects, only the matter has to be protected against deterioration to secure their survival, for more complex artefacts, documents also have to be preserved that ensure their readability and comprehensibility. The practice of writing manuals spans a history of several hundred years. A text from the Middle Ages, for instance, describes the procedure of how to probe the depth of a sea or river with the instrument of the astrolabe. The explanations given only partially work, because apparently reliable and unreliable sources were mixed. Authors that only copy the descriptions of others without having used the device endanger the transmission of the vital information by adulterating it.[242] Anders accentuates the creation of durability by "platonisation," but the fact that these objects need to be understood to survive actually renders them very fragile.

During its lifetime, the performance of the material setup visibly resolves its mysteriousness, but when it breaks or becomes obsolete and stops, the enigma emerges. Intellectual and practical effort is then necessary to reconstruct it. The longer it takes for this to happen, the more the unreadability increases, up to a point where even its former overall purpose is forgotten, either because its documentation was lost, or because it was ripped out of a larger machine context, or because its material basis was mutilated. The artefact gets into this peculiar situation of Odradek-like mysteriousness, because it is not sufficiently charged by interest in it.[243] Being purposeless and no longer understood, the thing loses its place in the world, the location where it was effectively used, and starts to wander around:

> "'Well, what's your name?' you ask him. 'Odradek,' he says. 'And where do you live?' 'No fixed abode,' he says and laughs; but it is only the kind of laughter that has no lungs behind it. It sounds rather like the rustling of fallen leaves."[244]

The obsolete artefact no longer worries inventors, engineers, or factory owners trying to make it as effective as possible, but only people that have the leisure to be troubled by its mysteriousness, like the house father. Many websites for enthusiasts of obsolete technology contain a section with photos of unidentified components asking the visitor for help with appeals like: "No idea what this fine piece of technology is. It has a number W75 309-1-A. Any comments and offers appreciated." (cf. Figure 4)[245]

241. Jakkrit Kuanpoth, The political economy of the TRIPS agreement, in: *Trading in Knowledge*, eds. Graham Dutfield, Ricardo Meléndez-Ortiz, and Christophe Bellmann (London, 2003), pp. 45–56, p. 46.

242. Arianna Borrelli, The flat sphere, in: *Variantology 2. On Deep Time Relations of Arts, Sciences and Technologies*, eds. S. Zielinski and D. Link (Cologne, 2006), pp. 145–166, especially pp. 148–151 and 164–166.

243. Slavoj Žižek in contrast interprets Odradek as "jouissance," because he is purposeless, and later, as "libido as organ." Slavoj Žižek, Odradek as a political category. *Lacanian Ink* 24/25 (2005): 136–155.

244. Kafka, Cares of a family man, p. 428.

245. Crowthorne Tubes website, http://crowthornetubes.com/.

Fig. 4: Enigmatic object from a website specialising in historic valves.

Reconstructing Odradek, I.

In a newspaper article from 1934 (the same year as Anders' lecture) commemorating the tenth anniversary of Kafka's death, the German philosopher Walter Benjamin wrote about "the most peculiar bastard in Kafka that the Prehistoric has conceived with the Guilt": "The attics are the location of the discarded, forgotten effects. [...] Odradek is the form that things assume in oblivion. They are disfigured."[246] In the last sentence, he uses the word "entstellt," which signifies, according to Grimm's German dictionary: "aus der rechten stelle, fuge oder gestalt bringen" (to get out of the right place, junction, or shape), while "fuge" is explained as "die enge verbindung zweier aneinander passender theile" (the narrow joint of two parts fitting to each other).[247] In German the word once meant to have been forced away from the right, the fitting location.

---

246. Walter Benjamin, Franz Kafka. Zur zehnten Wiederkehr seines Todestages [1934], in: W. Benjamin, *Gesammelte Schriften*, vol. II 2 (Frankfurt a. M., 1977), pp. 409–438, quotation p. 431: "Der sonderbarste Bastard, den die Vorwelt bei Kafka mit der Schuld gezeugt hat, ist Odradek. [...] Die Böden sind der Ort der ausrangierten, vergessenen Effekten. [...] Odradek ist die Form, die die Dinge in der Vergessenheit annehmen. Sie sind entstellt." English translation: W. Benjamin, Franz Kafka: On the tenth anniversary of his death, in: *Illuminations*, trans. Harry Zohn (New York, 1969), pp. 111–140, quotation p. 127.

247. Jacob and Wilhelm Grimm, *Deutsches Wörterbuch* (Frankfurt a. M., 2004), entries "entstellen" and "fuge."

The broken construction is reminiscent of the educational children's game "spool racer." The experimenter inserts a rubber band through the hole in a spool, fixes it with a toothpick, and then winds it up with a pencil at the other end. Releasing the device results in the fast rotation of the spool, which sets the construction into rapid and rather uncontrolled motion.[248] If we consider Odradek as an obsolete artefact that has lost its place and now wanders around as an enigma, we could speculate that a part of it is missing and renders it unreadable. If another rod were added to the crossbar that sticks out of the spool, it already starts to look more useful (cf. Figure 5).[249]

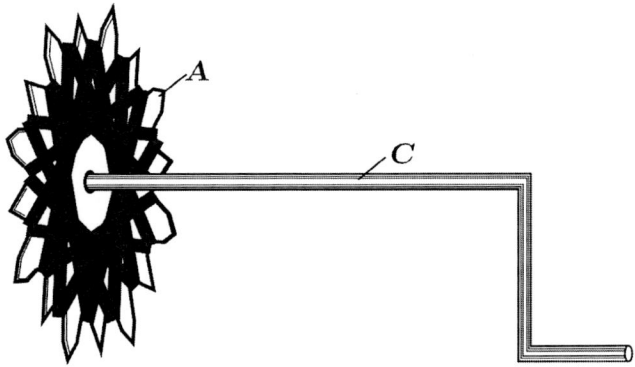

Fig. 5: A speculative attempt at reconstructing Kafka's Odradek; A: star-shaped spool, C: crank lever.

The crank lever could indicate that the artefact was used in the context of a bigger mechanical assembly. Around thirty years before Kafka wrote his story, in 1887, the British engineer John Keats patented an "apparatus for winding thread." At that time, the biggest challenge facing shoe manufacturers in England was to mechanise their production to be able to compete internationally.[250] While many of his guild emphatically opposed the new developments, the poet's namesake took up the challenge and in 1864 solved one of the most difficult problems, the mechanical fastening of a sole to an upper. He invented a specialised sewing apparatus, the "Crispin," in collaboration with the owner of a big shoe company in Street, Somerset, William S. Clark of C. & J. Clark, which still exists today.[251]

---

248. Cf. John Graham, *Hands-on Science. Forces and Motion* (New York, 2001), p. 8.

249. The author wishes to thank the cobbler Willi Schiffer, Cologne, for this suggestion. Personal communication, 25 February 2007.

250. Cf. George B. Sutton, The marketing of ready-made footwear in the nineteenth century. A study of the firm of C.&J. Clark, in: *Capital, Entrepreneurs and Profits*, ed. Richard P.T. Davenport-Hines (London, 1990), pp. 41–60. For the full history of the company, cf. G.B. Sutton, *C. & J. Clark 1833–1903. A History of Shoemaking in Street, Somerset* (York, 1979), especially pp. 131–167.

251. Boot-making machinery, in: *Spon's Dictionary of Engineering*, eds. Edward Spon and Oliver Byrne (Oxford, 1869), vol. 1, pp. 485–498, especially p. 495; and Sutton, *C. & J. Clark*, pp. 145–148. The corresponding patent is: John Keats and William S. Clark, Improvement in sewing-machines (Patent no. US 50,995, filed in

In the 1870s, Keats devised a machine for "twisting silk thread," which was built by the engineering manufacturer Greenwood & Batley in Leeds and exhibited at the International Exhibition held in Vienna in 1873.[252] He then turned to the mechanically complicated problem of winding up yarn and conceptualised and improved the necessary apparatus in the six years between 1887 and 1893 (cf. Figure 6).[253] He died around 1902, as documented by several posthumous patents applied for by his son.[254]

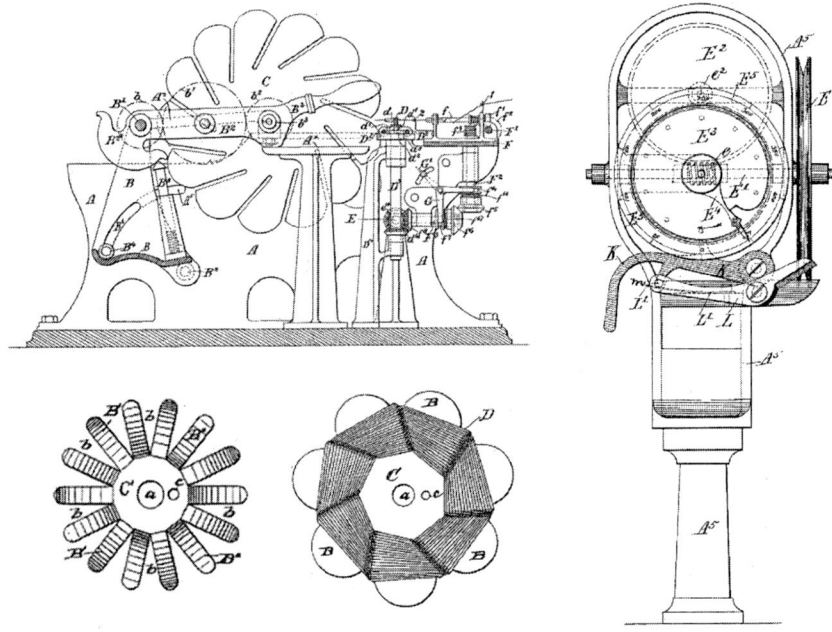

Fig. 6: Keats' winding machine, 1887, its cutting mechanism, and two of his star-shaped holders.

Keats' "apparatus for winding thread" was driven manually by a crank handle and shifted the yarn from left to right while sliding it through the slots of a rotating, star-shaped disk, "the laps of thread being made alternately on

England on 11 October 1864). All shoes sold by Clarks today are manufactured in the Far East.

252. Keats' silk thread twisting machine. *Engineering Weekly* (22 January 1875): 61 and 63.

253. J. Keats, Apparatus for winding thread (Patent no. US 440,811, filed in England on 12 January 1887); J. Keats, Holder for thread (Patent no. US 440,812, filed in England on 12 January 1887); J. Keats, Machine for winding thread upon disk-holders (Patent no. US 456,671, filed 3 September 1890); J. Keats, Machine for winding thread upon star-shaped holders (Patent no. US 543,185, filed in England on 14 January 1893). Cf. H. Glafey, Maschine zum Aufwickeln von Fäden auf Garnhalter mit sternförmigen Armen. *Dingler's Polytechnisches Journal* 269 (1888): 248–250, and H. Glafey, Maschinen zum Aufwickeln von Fäden auf Garnhalter mit sternförmigen Armen. *Dingler's Polytechnisches Journal* 285 (1892): 221–225.

254. J. Keats, Machine for preparing the soles of boots or shoes (Patent no. US 742,444, applied for by John Charles Keats, 26 September 1902).

opposite sides of the thread-holder and upon each spur or arm in succession, or on two spurs or arms in succession."[255] In the second version of the machine, the size of the spool ("five, seven, nine, or eleven spurs") and the kind of winding could be controlled by a gearshift. The quantity of thread on the disk was measured and when a certain length had been reached, the apparatus automatically cut it and switched off. Winding thread in this way was much easier than with round spools, because the yarn did not need to be guided continuously up and down the body of the roll. Keats wrote that the "holder of novel construction" would, moreover, "display the thread in a better manner" and "present advantages in unwinding not hitherto possessed by thread-holders," because the thread would "run off in the plane of rotation of the holder, and that in a smooth and uniform manner, the only deviation from a straight line being due to the thickness of the material of which the arms of the holder are made."[256] The star-shaped disks that he preferably produced from "glazed card board," but also mentioned "wood, or metal sheets" as possible materials, later proved useful for soldiers so they could repair their uniforms (and the flesh wounds under it) in the field; they could not carry spools on them that might break and possibly injure them when jumping on the ground.[257]

That the crossbar is made of wood in Kafka's story may indicate that either the object belongs to machinery of an even earlier era, or that the author tried to prevent it being interpreted as an obsolete artefact or part of apparatus from the past. The remark in the paragraph following the one cited above points in this direction, but Odradek's agility prevents any definite conclusion:

"One is tempted to believe that the creature once had some sort of intelligible shape and is now only a broken-down remnant. Yet this does not seem to be the case; at least there is no sign of it; nowhere is there an unfinished or broken surface to suggest anything of the kind; the whole thing looks senseless enough, but in its own way perfectly finished. In any case, closer scrutiny is impossible, since Odradek is extraordinarily nimble and can never be laid hold of."[258]

That no sign of breakage can be observed is consistent with the above interpretation, since it would be hidden under its right leg. The case that the enigmatic association of material components has never existed will be covered below.

In Kafka's time, around 1915, the technique and form of the thread star reappeared in the context of the cheap and mostly home-made crystal radio receivers, consisting only of an antenna, a coil and an earphone, without any power supply. In the infancy of the medium of ham radio, enthusiasts tried to ameliorate the reception of the AM waves by devising spools in a large variety of shapes. "Honey comb" and "spider web" coils, based on the form of a star and

---

255. Keats, Machine for winding thread upon star-shaped holders, p. 1.
256. Keats, Holder for thread, p. 1f.
257. The author is indebted to Rudi Wache, Amann Sewing Threads GmbH Bönnigheim, for this information. Personal communication, 1 March 2007.
258. Kafka, Cares of a family man, p. 428 (Translation slightly altered, D.L.).

wound similar to those in Keats' invention, proved to be very effective in this respect and were widely adopted (cf. Figure 7).[259] From 1909 onward, Kafka's fiancé Felice Bauer worked at the Carl Lindström A.G. in Berlin and marketed the "Parlograph," an early dictating device that recorded on wax cylinders.[260] In 1924, the year of the writer's death, the company introduced its first radio, the G1.

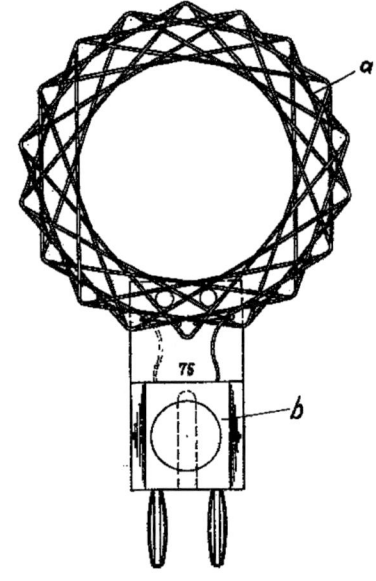

Fig. 7: Star-shaped spool for radio, Carl Gamke, 1927.

While application areas, implementations, and concrete objects pass away, certain basic techniques or *schematic forms* are put into practice again and again.[261] The obsolete, enigmatic artefact Odradek keeps returning:

"Often for months on end he is not to be seen; […] but he always comes faithfully back to our house again."[262]

The winding method of these antennas and of the thread star originated from the craft of basketry; more specifically, from procedures like the triple twist, where the weavers are placed between three consecutive spokes and are brought in succession over two and under one of them.[263] The knowledge of technical *schemata* incorporated into the artefact can be transcribed into different inventions in other contexts after it has become useless. Sohn-Rethel observed a spectacular instance of such a transfer: "A further example would be the wheel-motor, which, liberated from the constraints of some smashed-up motorbike, and revolving around a slightly eccentric axis, whips the cream in a latteria." His analysis was that for the Italian "the essence of technology lies in

---

259. They were patented in Germany by Carl Gamke, Flachspule nach Art einer Korbspule (Patent no. DE 484,261, 15 February 1927).

260. Sander L. Gilman, *Franz Kafka* (London, 2005), p. 61.

261. Cf. I. Kant, Of the schematism of the pure concepts of the understanding, in: *Critique of Pure Reason* [1781], trans. Norman K. Smith (London, 1933), A137–A147, A138: "How, then, is the *subsumption* of intuitions under pure concepts, the *application* of a category to appearances, possible? […] Obviously there must be some third thing, which is homogeneous on the one hand with the category, and on the other hand with the appearance, and which thus makes the application of the former to the latter possible. This mediating representation must be pure, that is, void of all empirical content, and yet at the same time, while it must in one respect be *intellectual*, it must in another be *sensible*. Such a representation is the *transcendental schema*."

262. Kafka, Cares of a family man, p. 428.

263. Cf. Anna A. Gill, *Practical Basketry* (Philadelphia, 1916), p. 29. Archaeological finds indicate it is older than pottery; the earliest objects are dated to approximately 10,000 BC; cf. George W. James, *Indian Basketry* (Kila, MT, 2005), p. 13: "There is no question that baskets preceded pottery making."

the functioning of what is broken" and that hence the artefact quickly oscillates between effectiveness and defect. It is mirrored by processes of technical recurrence, or reincarnation, which span enormous periods of time.[264]

In a theoretically momentous report for the "Conseil des Universités" of Quebec in 1979, the French philosopher Jean-François Lyotard denied the possibility of general and coherent theoretical accounts altogether: "The grand narrative has lost its credibility, regardless of what mode of unification it uses, regardless of whether it is a speculative narrative or a narrative of emancipation."[265] That this proposition is without substance can already be seen from the fact that it is paradoxical (in a weak sense). If all unifying accounts have lost their credibility, then so has Lyotard's very general one about delegitimation. Logically, it follows that some are exempt from this rule, as in the classical case of the lying Cretan. The history of ideas has to continue to tell "grand narratives" to reduce complexity, otherwise it turns into an extremely voluminous, unordered list of facts. "A theory is good only to the extent that it compresses the data into a much smaller set of theoretical assumptions and rules for deduction," writes Gregory Chaitin in an article on uncertainty in mathematics, a discipline shaken by crisis much more than the "soft" sciences, because it possesses methods of falsification.[266] Even if Chaitin's formal requests cannot be fulfilled in the history of ideas, one of the more stable fundaments it can rely on in its quest for lossless reduction of complexity are technical artefacts, because these materialised thoughts travel with slower relative speed in the stream of time than the narratives that conceive and accompany them. Unlike hermeneutic fiction and free association, their identity over time can be proven, and their possible functionalities can be experienced and experimented with at will.

The science historian Hans-Jörg Rheinberger has summarised this research strategy with admirable clarity:

"My emphasis is on the materialities of research. [...] My approach tries to escape this 'theory first' type of philosophy of science [like underlining the theory-ladenness of observation in the footsteps of Lyotard]. For want of a better term, the approach I am pursuing might be called 'pragmatogonic.'" "Instead of reading a history of objectivity from concepts, I embark on reading a history of *objecticity* from material traces."[267]

Historical materialism is "turned off its head, on which it was standing before, and placed upon its feet."[268]

264. Sohn-Rethel, Ideal des Kaputten, p. 37 (Translation altered, D.L.).

265. Jean-François Lyotard, *The Postmodern Condition: a Report on Knowledge* [1979] (Manchester, 1984), p. 37.

266. Gregory J. Chaitin, Computers, paradoxes and the foundations of mathematics. *American Scientist* 90. 2 (2002): 164–171, quotation p. 170.

267. Hans-Jörg Rheinberger, *Toward a History of Epistemic Things. Synthesizing Proteins in the Test Tube* (Stanford, 1997), pp. 26 and 4.

268. Friedrich Engels, *Ludwig Feuerbach and the Outcome of Classical German Philosophy* [1886] (New York, 1941), p. 44.

Reconstructing Odradek, II.

The second thought experiment in reconstructing the technical schemata Kafka was copying for Odradek starts with another close reading of the only remaining information: several paragraphs written not by the inventor of the device, but by an external observer. The rod leaves the star-shaped spool in the middle, but we probably deduced the setup of Figure 1 too early, and should have included in the possible configurations the one shown in Figure 8, which was created by turning the wheel 90 degrees around the y-axis.

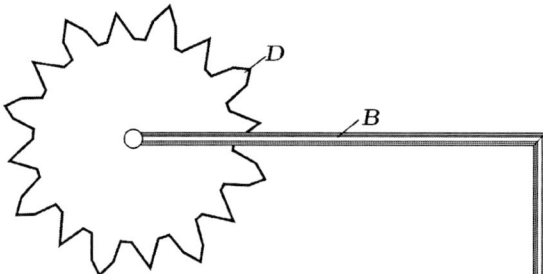

Fig. 8: Another speculative attempt at reconstruction of Kafka's Odradek; B: supporting bar, D: star-shaped disk.

It is very likely that Kafka was introduced to such a device made of iron when he attended the secondary school "Altstädter Deutsches Gymnasium" in Prague from 1893 to 1901. The English mathematician and natural philosopher Peter Barlow devised it in 1822 within the context of experiments on electromagnetism, and demonstrated that current could produce rotary motion (cf. Figure 9).[269] He mounted a star-shaped wheel on an axis above a trough of mercury in such a way that its points made contact with the electrically conductive liquid. When he applied current, it flowed along the radius of the disk and produced a field that interacted with a permanent or electric U-magnet beneath it. "In effect, the points of the wheel constituted a series of wires radiating from the axle."[270] The star-shaped disk started to revolve.

269. It was first mentioned in a letter to Michael Faraday on 14 March 1822, cf. *The Correspondence of Michael Faraday*, ed. Frank A.J.L. James (London, 1999), p. 254f.

270. Thomas B. Greenslade, Jr., Apparatus for natural philosophy: Barlow's wheel. *Rittenhouse* 1 (1986): 25–28, p. 27. Cf. Daniel Davis, *Davis's Manual of Magnetism* (Boston, 1852; online: http://quod.lib.umich.edu/m/moa/ajn7462.0001.001), pp. 103–115. Cf. L. Pearce Williams, ed., *The Selected Correspondence of Michael Faraday, Vol. 1: 1812–1848* (Cambridge, 1971), p. 132f. My thanks go to John D. Jenkins for providing the high-resolution image of the wheel shown at the Spark Museum, http://www.sparkmuseum.com/MOTORS.HTM.

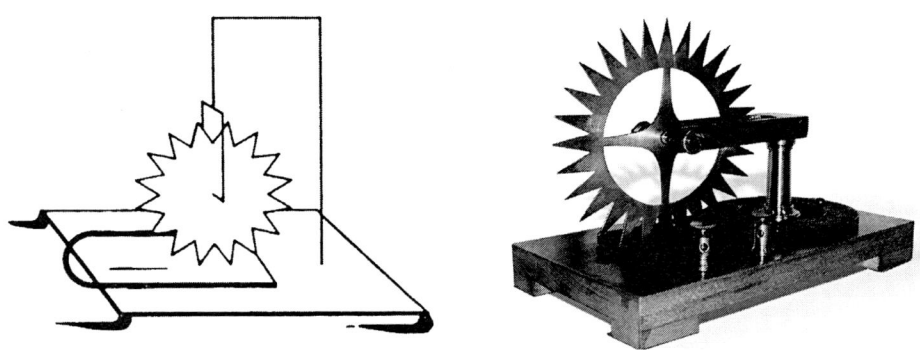

Fig. 9: Barlow's wheel, original drawing, 1822, and demonstration apparatus, ca. 1890.

While the invention demonstrated the possibility of electro-mechanical motion, it did not realise it effectively enough to be useful as a driving force for machinery. The final setup presumably emerged from a series of experiments with earlier prototypes, whose components and forms we do not know. A direct predecessor, however, is known: Michael Faraday's revolving wires from 1821.[271] He suspended an electric filament next to a permanent magnet in a cup of mercury, and when he applied current to it, due to the field generated, it started to rotate. Barlow's wheel was a popular object for demonstrations in physics classes and several companies continued to produce it even after it had been superseded, well into the twentieth century. Further developments by several inventors in the nineteenth century turned the tinkered, experimental setup into a most useful and, after some time, also robust artefact: the electrical motor, which allowed things to move around "nimbly" like Odradek, when it was used to drive their wheels. The fragile association produced by a *bricoleur* provided functions that could be used in a variety of contexts outside the laboratory, and was consequently transformed into a unified component.[272] Brushes replaced the impractical trough of mercury and applied engineers hardened the device to withstand the randomness of ignorant widespread use. The resulting reliable technical element could then be used, in turn, as a part in a newly invented, tinkered setup, to create effects more interesting than what had by then become the standardised electric motor.

Figure 10 shows three experimental artefacts from approximately the same period that never became effective outside the laboratory – or probably not yet.[273]

271. Michael Faraday, On some new electromagnetical motions, and on the theory of magnetism. *Quarterly Journal of Science and Art* 12 (1821): 74–96.

272. Cf. H.-J. Rheinberger, From experimental systems to cultures of experimentation, in: *Concepts, Theories, and Rationality in the Biological Sciences*, eds. Gereon Wolters and James G. Lennox (Konstanz, 1995), pp. 107–122, p. 111: "Scientists are *bricoleurs*, not engineers."

273. The reader is invited to guess at their purpose.

Fig. 10: Demonstration apparatus from the nineteenth century.

The line of very early computers developed at the University and the firm of Ferranti in Manchester from 1948 to 1951 serves as an example of what happens when complex idealised machinery becomes obsolete. If it is reasonable to regard a group of engineers that deploys the possibilities of a certain technology as a "culture," united by a set of objects, theories, and methods, then the title of a once popular fiction story might be suitable for the discussion.[274]

The Last of the Mohicans

The electrical engineer Frederic C. Williams (1911–1977) developed the first reliable means of volatile, random-access memory, the Williams tube, at Manchester University in 1946 (cf. Figure 11). His assistant, Tom Kilburn (1921–2001), worked out most of the intricate technical details.[275] On the screen of a standard cathode-ray tube (CRT) 1280 dots were detected by a pickup-plate in front and regenerated before they faded away. At first, the engineers manually switched the bright points on and off with a keyboard.[276] To test if their store would also work reliably when operated quickly and in a mode of random access, they built a very simple computer, but one that was Turing-complete, around it, the Manchester "Baby" prototype (1948). It evolved organically into the Manchester Mark I (1949), which was then industrially produced as the Ferranti Mark I (1951), two of which were delivered to the Computer Departments at Manchester and Toronto University.[277]

274. Cf. James F. Cooper, *The Last of the Mohicans* (Philadelphia, 1826).

275. Frederic C. Williams, Improvements in or relating to electrical information-storing means (Patent no. GB 645,691, 11 December 1946).

276. F.C. Williams and Tom Kilburn, A storage system for use with binary-digital computing machines. *Proceedings of the Institution of Electrical Engineers Pt. II* 96 (March 1949): 183–202, here p. 193.

277. For the emergence of the storage technique out of radar research and a discussion of early programming on these machines, see Chapter 3.

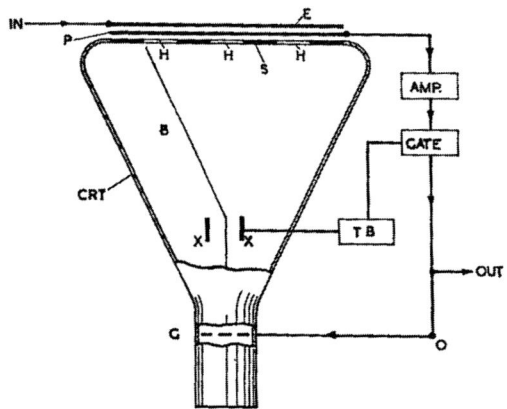

Fig. 11: Williams tube, 1946.

When the first machine using the volatile memory was built in 1951, the feet of those who would carry out the newly invented technology were already at the door.[278] Core store had been developed in 1949 by An Wang from Harvard University and was incorporated into the flight simulator project "Whirlwind" at the Massachusetts Institute of Technology (MIT) by Jay Forrester (cf. Figure 12). From 1954 on, it was commercially available in the form of IBM's 737 magnetic core unit, and by the mid 1950s, most major computer manufacturers had switched to the new technology.[279] At the end of the decade, industrial plants set up in the Far East lowered the cost of the component considerably.[280] In 1954, the team under Tom Kilburn completed the next computer at Manchester University, "MEG," employing Williams tubes for storage. Before producing it and marketing it as "Mercury" in 1957, the Ferranti company replaced them with core memory. One of these machines was installed in the Computing Laboratory of the University in the same year and superseded the Ferranti Mark I, which was dismantled in June 1959.[281] The culture of electrical engineers modifying CRTs in intricate ways with the purpose of storing data in them, their technological progress and insights into the properties of the material setup did not last; it only endured for a brief period of eight years.

278. Cf. Georg W.F. Hegel, *The Phenomenology of Mind*, trans. James B. Baillie (Mineola, 2003), p. 43, and Acts 5:9: "Behold, the feet of them which have buried thy husband are at the door, and shall carry thee out."

279. The US American IBM 701, announced 1952, and the 702 a year later used Williams tubes, while the 704 in 1954 employed core memory; the American UNIVAC 1103 from 1953 operated on Williams tubes, the 1103A in 1956 on core memory; so did the Edsac II from 1957 built in Cambridge, UK, and the American Illiac II from 1962, which had relied on mercury delay lines before. The latest integration of Williams tubes was in the Russian Strela-1 computer, which was produced between 1953 and 1956 at the Moscow Plant of Computing–Analytical Machines. Cf. Edwin D. Reilly, ed., *Concise Encyclopedia of Computer Science* (Chichester, 2004), p. 262ff., and Simon Lavington, *Early British Computers* (Manchester, 1980), pp. 31–35.

280. Cf. Robert Slater, *Portraits in Silicon* (Cambridge, MA, 1987), p. 97.

281. Martin Campbell-Kelly, Programming the Mark I: Early programming activity at the University of Manchester. *Annals of the History of Computing* 2. 2 (1980): 130–168, especially p. 131.

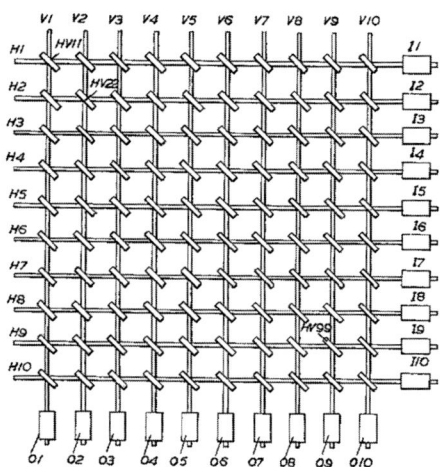

Fig. 12: Core memory, Wright 1951 (Patent no. US 2,667,542).

After that, the originator of the idea to use radar tubes as volatile memory, Frederic Williams, worked on other components, mainly electrical motors. He stayed in the Electrical Engineering Department when the Computer Science Department broke away from it in 1964. The newly founded institution was headed by his colleague Tom Kilburn and developed the next few generations of computers, adapting to the successive waves of new technology: MEG/Mercury from 1951 to 1954, in parallel with the Transistor Computer, MUSE/Atlas from 1956 to 1962 (built from core store and transistors), and MU5 from 1966 to 1974 (based on integrated circuits and core store).[282] Every time a new component was invented, the knowledge of the *bricoleurs* was rendered obsolete: "For engineers, the juggernaut of ever-changing technology presents a constant problem of learning new things and a reluctance to forget the old, in case the old may turn out to be useful."[283] If they can no longer keep up with the speed of developments, they are regarded by those around them as the last of an almost extinct exotic tribe, as Mohicans, with a mixture of respect and pity. Kilburn retired in 1981 and died in 2001.[284]

Every time the Computer Science Department built a new machine, the old ones were disassembled, so their parts could be re-used in other projects. This is why only singular components of the Ferranti Mark I remain, like several of the double Williams tubes, several logic doors, and possibly the original 19"

---

282. The details can be found in S. Lavington, *A History of Manchester Computers* (Manchester, 1975).

283. Christopher P. Burton, Replicating the Manchester Baby. Motives, methods, and messages from the past. *Annals of the History of Computing* 27. 3 (2005): 44–60, p. 44.

284. Cf. Brian Napper, Tom Kilburn (1921–2001); online: http://curation.cs.manchester.ac.uk/computer50/www.computer50.org/mark1/kilburn.html.

steel rack of the prototype.[285] Earlier computing apparatus suffered a similar fate, albeit for other reasons. The code-breaking machinery of the British, "Colossus," was ordered to be destroyed into "pieces no bigger than a man's hand" by Churchill after the war, some parts being secretly taken over to Manchester University's Electrical Engineering Department (some of the electronic panels and probably a teleprinter and power supplies).[286] Its US counterpart, the special-purpose ballistic calculator ENIAC, was switched off in 1955 and taken apart; portions of it are on display at several locations on the continent.[287]

Along with the "Last of the Mohicans," Tom Kilburn and a few other electrical engineers directly involved in the development, the complete and living knowledge of an obsolete generation of automata, which can only be acquired by building them, passed away, and with it, the possibility of obtaining details from them that were transferred informally and undocumented, either because they were presupposed as "common knowledge" or because they were considered profane. When writing an emulator of the Ferranti Mark I to run Christopher Strachey's Love Letter Program from 1952, the author was unable to determine the functions of the buttons and switches on its console because the available documentation did not cover them at all.[288] Fortunately, a retired computer scientist from Manchester University, Brian Napper, found "an engineer-biased description written in 1953 for ARE" (the Atomic Research Establishment) in his archive specifying their use.[289] In the end the durability of the paper on which the copies of the manuals circulate, as well as the interest of the owners in their conservation and their readiness at transcription determine the duration of survival of the remaining information and, hence, of the artefact itself.

The ingenious German computer pioneer Konrad Zuse (1910–1995), who invented the very first high-level programming language "Plankalkül" around 1945 and built several early machines, could also have served as an example of the "Last of the Mohicans" of an even earlier technical era.[290] The Z1 and Z2 from 1936 and 1939 used mechanical switches for storage and calculation, which

---

285. Material being re-used "cost nothing, and was known to work in a well-understood way so reduced learning time." C.P. Burton, personal communication, 8 March 2007.

286. Cf. B. Jack Copeland, A Brief History of Computing; online: http://www.alanturing.net/turing_archive/pages/Reference%20Articles/BriefHistofComp.html: "Some of the electronic panels ended up at Newman's Computing Machine Laboratory in Manchester […], all trace of their original use having been removed." Campbell-Kelly, Programming the Mark I, p. 135: "The teleprinter equipment was acquired by Turing through contacts with the Foreign Office Department of Communications at Bletchley Park." Burton, Replicating the Baby, p. 53: "Indeed in 1948 Williams may have obtained [power] units indirectly from dismantled Colossi."

287. Parts are exhibited at the School of Engineering and Applied Science at the University of Pennsylvania, the National Museum of American History in Washington, DC, and elsewhere. Cf. Ulf Hashagen, ed., History of Computing: Software Issues (Berlin, 2002), p. 265.

288. Cf. Chapter 3.

289. B. Napper, personal communication, 10 May 2006. Brian Napper sadly died in 2009.

290. The first published account of "Plankalkül," developed purely theoretically, dates from 1948: Konrad Zuse, Über den allgemeinen Plankalkül als Mittel zur Formulierung schematisch-kombinativer Aufgaben. Archiv der Mathematik 1. 6 (1948): 441–449.

moved metal pins into one of two positions. From 1940 to 1955 Zuse employed electro-mechanical relays (Z3, Z4, Z5, Z11). Already in 1937, his friend and fellow worker Helmut Schreyer, who wrote his dissertation about "tube relays" at the "Institut für Schwingungsforschung" (Institute for Oscillation Research) in Berlin under Wilhelm Stäblein, advised him to use thermionic valves, but Zuse considered it to be just "one of his wild ideas."[291] Since his developments were considered strategically unimportant by the German government, Zuse did not obtain the necessary funding and continued to use the mechanical switches and telephone relays, largely collected from discarded stock. Only in 1957 did the Zuse KG switch to vacuum tube technology and built the first computer on this basis, the Z22. All his early machines (Z1, Z2 and Z3) were destroyed in Allied bombardments.

The Rustling of Fallen Leaves

If machines only survive on paper, a general effect of abstraction can be observed, in which an increasing number of details is forgotten and material elements are sacrificed to the mysterious force of entropy. Full information is only stored in the artefact, as long as it is operational, and in the brains of its constructors, while they remember it. The intact apparatus is the shortest expression of its complexity.[292] Repairing it is therefore easier than understanding the mechanism completely, because only the functionality of the broken part needs to be reconstructed, while the rest of the knowledge is still implemented in hardware.

The surviving documentation can be roughly divided into several categories. The efforts of the next generation of electrical engineers in Manchester to resurrect the prototype computer, headed by Christopher P. Burton, serves to illustrate their usefulness and the information contained within.

Apart from material remains, construction plans of parts or even of the whole machine would permit one to form a precise picture of the hardware. Unfortunately, these only very rarely exist for pioneering efforts, because they are developed experimentally, by modifying or exchanging components and by trying to tune them to a point where their consonance produces the desired results:

> "The pioneers had no need for formal engineering drawings. Their working documents were a set of hand-drawn schematic circuit diagrams on a table (jokingly called Tom Kilburn's office), in the corner of the laboratory, together with their personal notebooks. […] The machine was constantly being modified and added to, as were the diagrams that always represented the current state. Those circuit diagrams no longer exist, so we have had to rely on secondary sources."[293]

---

291. Cf. Helmut Schreyer, *Das Röhrenrelais und seine Schaltungstechnik* (Ph.D. diss., Technical University Berlin, 1941; online: http://zuse.zib.de/item/EY9Ei3eiNg5WLx44). K. Zuse, *The Computer – My Life* (Berlin, 1993), p. 38. Six years later in Great Britain, the engineer Thomas Flowers encountered a very similar reaction when he proposed to build a code-breaking machine from around 1,500 electron tubes – Colossus.

292. Cf. Andrey N. Kolmogorov, Three approaches to the quantitative definition of information. *Problems of Information Transmission* 1. 1 (1965): 1–7.

293. Burton, Replicating the Baby, p. 50.

Engineers sometimes recorded invaluable technical details in their notebooks, as an aid to memorise certain aspects of the work under construction. This was the case in the rebuild of the Manchester computer, which could rely on such sketches of Dai Edwards, Geoff Tootill, and Alec Robinson, who copied some of the diagrams on the table. Documentation often occurs when a transfer of knowledge is involved. In this case, the three research students were trying to catch up with the work underway.

Secondly, patent applications usually contain the technical specifications necessary to protect the intellectual property of the invention, but consciously exclude numerous details, like the exact models or values of the parts used (cf. Figure 13). They only depict the device in a general way, and experimentation is required to actually build it. This "abstraction by concurrence" can be regarded as typical for the way an economic system based on competition prevents the dissemination of information. The transformation of the image part of patents from construction plans to circuit diagrams results from the idealisation of artefacts discussed earlier. Because the components look similar and do not reveal their functioning, which happens invisibly, the picture does not show the elements themselves, but represents their general function in abstract symbols: to switch current, to conduct it in only one direction, to resist, etc.

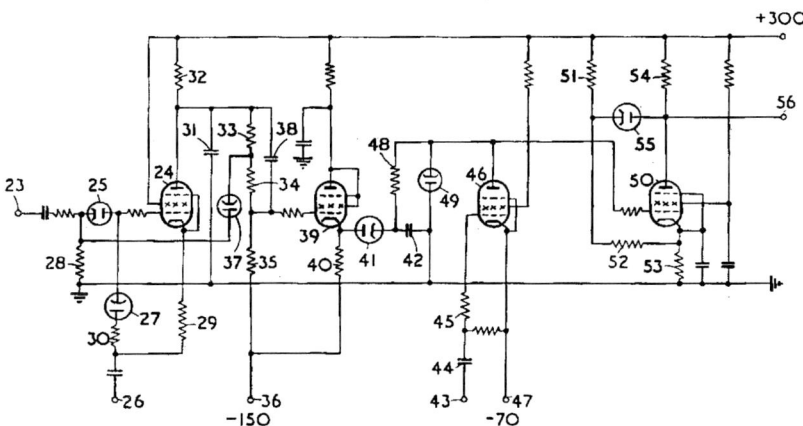

Fig. 13: Circuit diagram from the patent for Williams' storage tube, 1946.

Thirdly, articles in journals on electrical engineering serve a similar purpose: to publish the results of research and to publicly declare the importance of the invention, not necessarily to enable others to build the apparatus (cf. Figure 14). They may still contain diagrams specifying more of the actual design, like the characteristic values of the components. Even if the engineers helped others to actually construct the artefact, the technical details may have been communicated more informally, in letters or personal communication, and consequently might not have been preserved in written form at all.

101

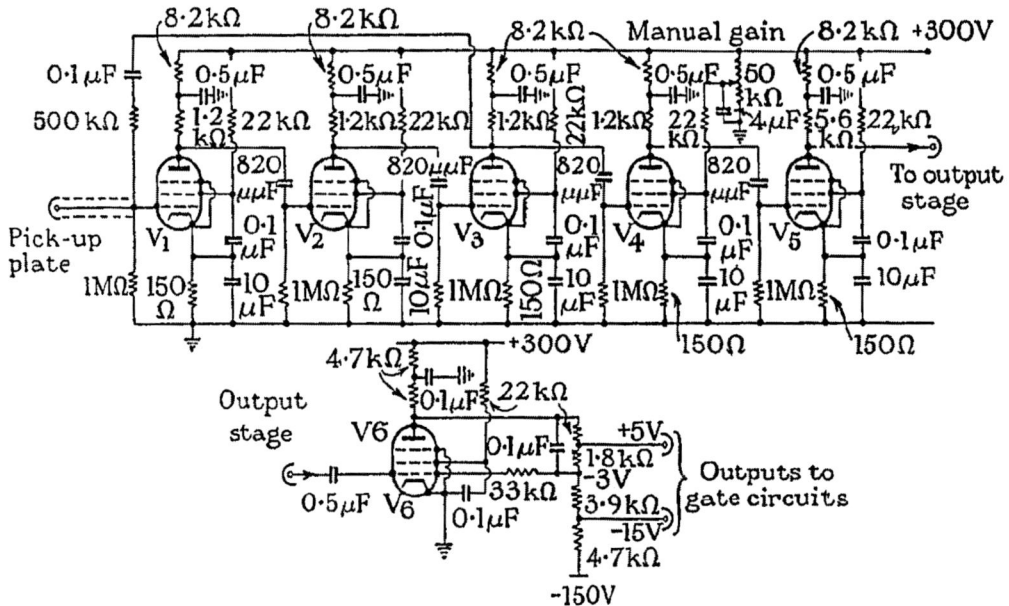

Fig. 14: Circuit diagram of the amplifier of the Williams tube from an engineering publication three years later.

Although the Baby rebuild project had acquired all of the surviving documentation and original components, their struggles to make the reconstructed Williams tubes work illustrates the effort required to turn an abstraction concrete again:

> "Although the description of the CRT memory operation given elsewhere seems straightforward, as an analogue electronic device it was **tricky to adjust**. Controls for brilliance, focus, astigmatism, high-voltage supply, deflection voltages, amplifier gain, threshold level, dash width, dot width, and strobe pulse timing **all interacted**. The secondary emission behaviour of the screen phosphor is not uniform in the early tubes [...] and furthermore they are susceptible to minute areas of zero emissivity, known as phoneys, where a bit cannot be stored."[294]

Fourthly, photographs might elucidate certain details that cannot be obtained elsewhere. In the case of the prototype reconstruction, photographs proved extremely useful, after they had been located: "The circuit diagram fragments [from the engineers' notebooks] and photographs were the key sources in our achieving authenticity." They made it possible to identify components that were not documented elsewhere, because they were not essential for the working of

---

294. Burton, Replicating the Baby, p. 56 (Emphasis mine, D.L.). Burton goes on to explain that "*Phoney* is a sort of slang word used by World War II pilots meaning a radar echo that was not a real target. The term carried over to the SSEM [Small Scale Experimental Machine] team in 1948." The meaning was inverted from an erroneously displayed dot that represented nothing (a "radar angel," in other words) to a memory location that could not display a dot; cf. Chapter 3.

the machine, like the exact type of push-button switch used on the keyboard: "I spent many hours gazing at the original photograph through a jeweller's eyeglass [...]. In one serendipitous moment, the pattern of holes on the panel suddenly brought to mind a set of five push-buttons that I had bought in 1953, and which I still had." Burton looked up the Royal Air Force part number in a catalogue, and finally found the exact component pictured on "a photograph of the Spitfire fighter's cockpit: It was a control box for the VHF radio."[295] Despite their idealisation and even as black boxes, the components possess a certain appearance that helps to tell them apart, even if their functionality can no longer be read off their surface.

Fifthly, manuals explained aspects of the machine to its future users to enable them to operate it.[296] They only contain the technical details necessary to understand its general functioning and do not permit it to be rebuilt. The engineering knowledge for its maintenance was usually transmitted by verbal instruction and training.

To be able to build a faithful replica of complex machinery, at least two types of documents are absolutely necessary. Diagrams of at least the most important subsections, which make it possible to reach an understanding of the logical functionality of the device. Photographs are needed to allow for the determination of the physical properties of the components, their concrete type and location within the whole. They also show objects like racks, cables and switches which are not specified in a circuit diagram.

Given the fact that a lot of essential information is often unavailable at the time the resurrection project is undertaken, the only means to fill in the gaps is by identifying with the engineers of the time, their technology and method of design, and by imagination, as indicated by Christopher P. Burton: "Lastly, by being immersed in the project I found it possible to 'think pioneer' and make plausible judgements as to what was correct."[297] Due to their interrelatedness, the information about the missing parts can be obtained from the remaining ones by deductive speculation, if somebody is willing to invest the time necessary to solve the puzzle. The situation corresponds to a rebus with known meaning in which one signifying element is missing and must be reconstructed. Although there are several possible objects that can fill in the gap, their number is limited, especially in the advanced technology of an era. It might well be that sometimes this method is misleading, but unfortunately and in a number of cases, it is the only chance to complete the missing parts in the picture of an extinct artefact.

---

295. Burton, Replicating the Baby, p. 51f.

296. These were only produced in Manchester after the prototype had evolved into a bigger machine; cf. Alan M. Turing, *Programmers' Handbook for the Manchester Electronic Computer Mark II* (112 page typescript, Manchester, 1951; online: http://www.turingarchive.org/browse.php/B/32).

297. C.P. Burton, personal communication, 8 March 2007. This kind of imagination is not to be confounded with free association and pure fantasy.

Inventions *in nuce*

Michael Thompson's *Rubbish Theory* is a witty study on the career of objects, which complements the linear arrows of economic theory that point from the producer to the consumer with the idea of a cyclical return. He distinguishes three categories into which things fall in different moments of their history (cf. Figure 15). In the first one, transience, the commodity circulates on the market and its value decreases slightly over time. Two factors, fashion, which moves in cycles, and technology, which proceeds linearly, bring about a more radical drop in its price. The latter also applies to singular components of a construction, as in the following example:

> "[I]n the mid-eighteenth century, the owner of a new house in the City of London would find that his property gradually became obsolete [...]. [H]is plumbing system discharging into a cesspit in his rear basement room, whilst perfectly adequate when the house was new, would gradually appear less and less attractive after the invention in 1779 of Alexander Cummings's patent water closet [...] [T]he march of technological evolution is irreversible and linear."[298]

While the water closet is still around today, 250 years later, more complex artefacts, especially in the early period after their invention, become obsolete and are replaced at a much faster pace, as demonstrated above. In a chapter aptly named "On the Duration of Machinery," Charles Babbage remarked in 1832: "During the great speculations in that trade [of making patent-net], the improvements succeeded each other so rapidly, that machines which had never been finished were abandoned in the hands of their makers, because new improvements had superseded their utility."[299]

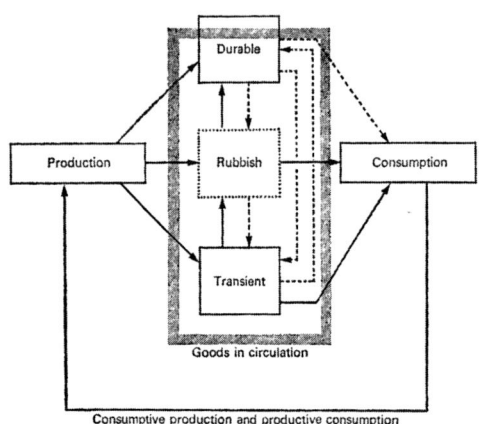

Fig. 15: Schematic view of Thompson's rubbish theory.[300]

When the commodity has become completely worthless, it enters a category considered covert by Thompson, that of rubbish. The object is suppressed and disappears from the consciousness of society. But astonishingly at a certain point, which is dependent on the end of its production, some artefacts resurface from the trash and are

---

298. Michael Thompson, *Rubbish Theory. The Creation and Destruction of Value* (Oxford, 1979), p. 38f.

299. Charles Babbage, *On the Economy of Machinery and Manufactures* (London, 1832), p. 286.

300. Full lines indicate possible, dotted ones impossible transfers. Thompson, *Rubbish Theory*, p. 113.

transferred to yet another category, durability. Eccentric evaluations by a few individuals lead to a renewed interest, enthusiasts start to collect them, and their value increases again. Finally, their price rises so high that they fall out of the market at its upper edge and are bought by museums, which reduce their relative speed in the stream of time to the smallest possible value.

This can be explained by the fact that its purposelessness turns obsolete technology from a complex means constructed by humans to achieve an end into a kind of natural beauty ("Naturschönheit," in German): "*Beauty* is an object's form of *purposiveness* insofar as it is perceived in the object *without the presentation of a purpose.*"[301] Already prior to its obsolesence, it can be interpreted aesthetically in this sense, because its inner workings – and thus the way in which it is realising its aim – are not easily understood by laypersons. The increasing disintegration of other copies of the same thing enhances the singularity and value of the remaining exemplars, and reverses the effects of Anders' "platonisation" (which, in eternalising its "idea," renders the concrete light bulb worthless) by fetishising the beautiful unicum.

Because Thompson takes artefacts as given without questioning how they come into existence, his theory does not treat another covert phase, in which they are first and foremost conceived, the experimental one, which precedes the overt state of transience. While belonging to this category, the material association is slightly altered; parts are replaced, optimised and tuned, until it delivers the desired performance. Before this happens, it possesses the same enigmatic, Odradek-like appearance as the one that belongs to the rubbish. Being no longer and not yet useful results in the same type of object. And indeed, the specific form might have been adopted from an obsolete artefact, or even from a device invented for an impossible purpose, like perpetual motion or communicating with the dead. In a number of cases, an experimental setup is devised for a certain aim, but since the function it provides cannot be used by a wider public, it stays in the closed realm of the laboratory for a shorter or longer duration, sometimes even centuries, without ever entering the state of transience.[302] Or its effects are purely epistemic in that they challenge a current paradigm as an anomaly and only have theoretical, but no immediate practical consequences.

By invalidating all attempts to understand Odradek as an obsolete artefact, Kafka ensures that the interpretations, which are always only partial, are ongoing. His riddle cannot be solved and, consequently, constitutes an enigma that survives. It is exactly this quality that Rheinberger takes as a condition for the durability of experimental systems, described aptly by Brian Rotman as "xenotext": "What [a xenotext] signifies is its capacity to further signify. Its value is determined by its ability to bring readings of itself into being. A xenotext thus has no ultimate 'meaning.'"[303] Because Kafka's story can be understood differently forever, but never perfectly, it lives on. Similarly, after their initial

---

301. I. Kant, *Critique of Judgement* [1790], trans. Werner Pluhar (Indianapolis, 1987), p. 84.

302. An example of this are letter wheels. After Alberti had established that they could be used as cryptographic devices in 1467, it took more than 450 years until the military employed them in this way; cf. Chapter 6.

303. Rheinberger, *History of Epistemic Things*, p. 36f.

purposes have become obsolete, it is possible for artefacts to reappear and be preserved as long as they allow a further reinterpretation of their material associations. It is impossible to determine beforehand if another fertile implementation is possible.[304]

The process of "poietic" engineering employed to reach a new understanding of the setup is similar to innovative processes in language. A word combines elements from one class, the alphabet, for the purpose of conveying meaning, and in ordinary usage convention prescribes which letters to use. The primary goal attained is symbolic, but would not be effective without imaginary (world reference) and real (contract, instruction etc.) consequences. Avant-garde poets create new significations by trying out new combinations of letters and words. The artefact, on the other hand, employs material elements from different classes (glass, air pump, electricity in the case of the light bulb) to achieve immediate material goals at first, but later also imaginary and symbolic ones (media, algorithmic machinery). The situation of designing an experimental setup corresponds to the free construction of a rebus for a given meaning. Forms need to be selected that are known to perform a certain function and must be combined in the right way. The number of possibilities is high to begin with but reduced by reasoning and experiment. Similar to poets, scientists challenge conventional implementations by modifying material associations to see if new effects arise.

The processes of replacing and tuning parts to make the *bricolage* work in the desired way can be compared to the operations of metaphor and metonymy in language. They exchange components or adjust certain parameters, but only slowly transform the formal core of the artefact. Rheinberger writes: "Science, viewed from a semiotic perspective, does not escape the constitutive texture of the inner workings of any symbol system: metaphoricity and metonymicity. Its activity consists in producing, in a space of representation, material metaphors and metonymies."[305] Poets are constrained in their innovative combinatorics by the linguistic conventions of society that secures understanding, and inventors and scientists by the laws of nature: "We have to recognise that the qualities objects have are conferred upon them by society itself and that nature […] plays only the supporting and negative role of rejecting those qualities that happen to be physically impossible."[306]

Certain transfers are theoretically ruled out by Thompson. An object cannot move from being rubbish to being transient or from durable to transient under normal circumstances governed by the laws of the market. The first one, he concedes, occurs "to a limited degree, which does not seriously threaten the boundary maintenance, in the business affairs of the dealer" and is "implicit in the slogans 'we want what you don't' and 'houses cleared free of charge.'"[307] He similarly shows that the second one only takes place in troubled times by quoting the

304. In parallel to Thompson's "father of all the Tiv," Odradek could be called the forgotten ancestor of all artefacts; Thompson, *Rubbish Theory*, pp. 65–69.

305. Rheinberger, Experimental systems, p. 114f.

306. Thompson, *Rubbish Theory*, p. 9.

307. Thompson, *Rubbish Theory*, p. 106.

example of wartime privations in Germany "that caused the (regretted) exchange of Old Masters for tins of corned beef."[308] In the construction of the artefact precisely these transfers happen on an immaterial level (cf. Figure 16).

The experimental phase reintegrates the outcome of earlier cycles. The components of the "machines for making the future"[309] can be of natural descent, like the bamboo filament of early light bulbs, but they usually result from prior inventions, like the vacuum tube. The transfer takes place from the transient category back into new, tentative systems. Techniques of associating and shaping components, technical schemata, can be extracted from museological or almost forgotten devices and can be transcribed into new, tinkered constellations. And it is the effects of artefacts as anomalies in basic research, changes of paradigms, which open up the possibility to invent new, vague ideas of purposes and their implementation. The transfer from rubbish or durable back to experimental and later transient transcends the demarcation lines between the categories postulated by Thompson. The "communicable things"[310] are transacted at each station of their life cycle, even if they no longer change their physical location, be it the rubbish dump or the museum.

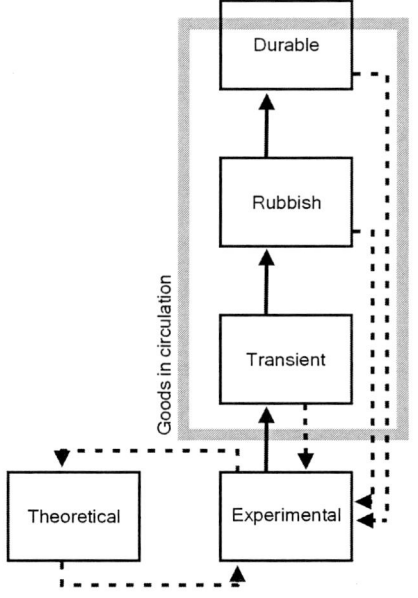

Fig. 16: A cyclical extension of Thompson's rubbish theory.[311]

The fact that epoch-making inventions started their life cycle as Odradeks, abstruse and purposeless creatures, which were sometimes doomed from the beginning, is a strong argument against forcing all research in science to be of an applied nature. This would have prevented the creation and transformation of absolutely useless artefacts that in the past led to important and far-reaching discoveries.

To raise some issues concerning algorithmic objects in particular, the last section discusses a case of crypto-history: the German Enigma and its decipherment by Polish mathematicians before and during the Second World War.[312]

---

308. Thompson, *Rubbish Theory*, p. 107.
309. Rheinberger, Experimental systems, p. 110.
310. Thompson, *Rubbish Theory*, p. 44.
311. Full lines indicate material, dotted ones immaterial transfers.
312. For the following see also the next chapter.

An Archaeology of Algorithmic Artefacts

If algorithmic machinery is described without experimenting with it, there is a danger of adulteration, even if the methodology employed is scientifically impeccable and the author's background knowledge very extensive.

On the first pages of *Seizing the Enigma*, the cryptohistorian David Kahn describes the *ethos* of his research: "The book [...] focuses upon personalities and rests as much as possible upon primary sources, namely documents and interviews." Kahn decided to include the technical details in his text: "This material [on Enigma cryptanalysis] may seem dry, but to leave it out would obscure a main point of this book: the fearful difficulty of the work of the cryptanalysts."[313] Because of the tactical relevance of the intelligence obtained, which played a role as important as the findings by radar and other technical advances, any account omitting the hidden cryptographic battle would be incomplete. The official historian of the British Crown, Harry Hinsley, estimated that the breaking of Enigma shortened the war by at least two years.[314] Given the prevalence of myths and cover stories concerning the events of World War II, it seems even more important than with other topics to rely as much as possible upon primary sources.

In *The Codebreakers*, Kahn gives an extensive and historically rich overview of cryptological technology throughout the centuries, and it is difficult to name any other author who possesses a comparably intimate knowledge of the field. However, when it comes to the first machines built by the Polish cryptanalysts against Enigma and the procedures executed on them, despite his cautious methodology, the book strangely teems with errors. Concerning the cyclometer, one of the earliest devices, Kahn writes: "The maximum of 26 [letters] was reached in only three ways, or cycles: two chains of 13; six chains of 10, 10, 2, 2, 1, and 1 letters each; and six chains of 9, 9, 3, 3, 1, and 1 letters each."[315] If this had been the case, the apparatus would have been useless, because it would not have reduced the number of combinatorial possibilities to a manageable level. Actually, 82 different characteristic patterns exist.[316] A later machine, the "bomba kryptologiczna," is described in the following way:

> "[The Polish cryptologist] Rejewski soon found a way of determining the keys when the indicators in three messages fulfilled certain conditions. [...] One pair [of the three pairs of Enigmas incorporated into the bomba] would be testing the indicators of messages 1 and 2, another those of messages 1 and 3, and the third those of messages 2 and 3."[317]

---

313. David Kahn, *Seizing the Enigma. The Race to Break the German U-Boat Codes, 1939–1943* [1991] (London, 1992), p. X.

314. B. Jack Copeland, Enigma, in: B.J. Copeland, ed., *The Essential Turing: Seminal Writings in Computing, Logic, Philosophy, Artificial Intelligence, and Artificial Life* (Oxford, 2004), pp. 217–264, p. 218.

315. Kahn, *Seizing the Enigma*, p. 63.

316. The author has programmed a simulator of the cyclometer, which can be found at http://www.alpha60. de/research/. It has been calculated that theoretically, 101 patterns are possible; cf. Kris Gaj and Arkadiusz Orłowski, Facts and myths of Enigma. Breaking stereotypes, in: *Eurocrypt 2003*, ed. Eli Biham (Berlin et al., 2003), pp. 106–122, p. 113.

317. Kahn, *Seizing the Enigma*, p. 73.

This is all the more astonishing since Kahn had access to the descriptions, and especially the appendixes, in the English translation of Kozaczuk's very detailed and precise book, which clearly describes that each pair was simply working on only one of the female indicators received.[318] It shows, and this is the only purpose of the two examples quoted, that one cannot understand, describe, or research these devices without experimenting practically with them, either by re-building them or by simulating them in software.[319] Turing's statement that "[i]t is possible to produce the effect of a computing machine by writing down a set of rules of procedure and asking a man to carry them out," his concept of a "paper machine," is theoretically and practically true for Enigma, but only theoretically (under the condition of almost eternal time) for the cyclometer and the bomba. The only reason these apparatuses were built is that it was no longer possible to perform the corresponding routines with "paper, pencil and rubber, and subject to strict discipline."[320] The concrete details of their functioning cannot be cap-tured in verbal descriptions, and even if there was a complete one, it would be impossible to grasp it without putting it into practice.

That these devices execute procedures that are too lengthy and complicated for humans to perform radically reinforces Kant's dictum that "we have com-plete insight only into what we can ourselves make and accomplish according to concepts."[321] Not incidentally, the mixing of theory and practice is already typical for the cryptanalyst's work. Rejewski wrote about the invention of the bomba and the Zygalski sheets: "Yet, within a very short time [...] we had two ideas, or rather, what was more important, we found ways to realise those ideas."[322] The historian or any other scholar trying to depict the developments suddenly finds himself in the same situation as the code-breakers. To verify the claims of different parties in the event and to really grasp these devices, one needs to be able to perform the procedures that have been employed.

It may seem that the title "computer history" is better suited to designate the investigation of algorithmic artefacts of bygone days. However, there are several arguments why the concept of archaeology should be preferred.

Because of the irrelevance, exchangeability, and opaqueness of their com-ponents, symbolic machinery moves with higher speed in the stream of time, quickly becomes obsolete, and is obliterated even faster. The language that is coupled to it to control it, like in Enigma, creates a new field in the usage

---

318. Marian Rejewski, Jak matematycy polscy rozszyfrowali Enigmę ("How Polish mathematicians broke the Enigma cipher"). *Wiadomości Matematyczne* 23. 1 (1980): 1–28. English translation: How the Polish mathe-maticians broke Enigma. Appendix D, in: Władysław Kozaczuk, *Enigma: How the German Machine Cipher Was Broken, and How It Was Read by the Allies in World War Two*, trans. Christopher Kasparek (Frederick, MD, 1984), pp. 246–271, p. 266f.

319. Kahn's book contains a wealth of details that are not found in other accounts and therefore remains an excellent historical survey on the topic.

320. A.M. Turing, Intelligent machinery (22 page typescript, report to the National Physical Laboratory, 1948; online: http://www.alanturing.net/turing_archive/archive/l/l32/L32-001.html), p. 5. The paper was first pub-lished 20 years later, in: Christopher R. Evans and Anthony D.J. Robertson, eds., *Cybernetics: Key Papers* (London, 1968), pp. 27–52.

321. Kant, *Critique of Judgement*, p. 264.

322. Rejewski, How the Polish mathematicians, p. 266.

of signs. After a very long time, the question of how to achieve certain effects through symbols could be posed in a different way. Unlike natural expressions, their complete arbitrariness is not conserved by collective tradition. In the proliferation of programming languages, which is a reaction to the ongoing software crisis, ways of formulating change in rapid succession and are forgotten.[323] The text in which Robert Recorde introduced the equal sign 450 years ago is understood today without much effort: "And to auoide the tediouse repetition of these woordes: is equalle to: I will sette as I doe often in woorke vse, a paire of parralles, or Gemowe lines of one lengthe, thus: ==, bicause noe. 2. thynges, can be moare equalle."[324] This does not apply to the words that caused the Manchester Mark I to compose love letters 45 years ago, even for programmers: "//I/ //ZO ZA/: DEQO AIQB RE/: S:LO DSWO IST/ ..."[325] Archaeology investigates buried, incomplete, and often enigmatic artefacts.

The study of algorithmic apparatuses opens up a terrain in which theoretical and practical aspects, immaterial procedures, and their technical implementation refer to and emerge from each other. The *difference of execution*, which separates them from other artefacts and consists in their autonomous carrying out of almost endless sequences of instructions, creates an opaque area between the input of the original data and the incalculable moment in which the machine returns with the result, which is not easily penetrated. The history of technical ideas cannot omit this time interval and is like cryptology plunged into the question: What exactly is executed here? The theoretical investigation is forced to pass through the practical work of reconstructing the apparatus and the operations performed on it. Not to proceed in this way would only leave the possibility of hermeneutically interpreting cryptograms based merely on their appearance, as Dadaist poems. The mixing of narration and practical reconstruction, of confirmed fragments and deducto-speculative complements is likewise typical for archaeology.

The display and output of an algorithm is a surface without obligation. The symbols that instruct the procedures can indeed be found in the source code. However, because of the fact that they are executed, reading them only provides a complete picture of the operations in the case of very simple algorithms. Only the execution unfolds the complexity implemented in them and allows the formulation of reliable propositions about processes that cannot be described or run on paper for principal and practical reasons. The authoring of experimental software becomes part of the theoretical investigation, because it is the only way algorithmic artefacts can be fully and concretely grasped.

---

323. Jean E. Sammet estimated in 1991 that over 1000 programming languages had been created in the USA alone since 1953. Only a few of them have survived to the present day; cf. J.E. Sammet, Some approaches to, and illustrations of, programming language history. *Annals of the History of Computing* 13. 1 (1991): 33–50, p. 48.

324. "Gemowe" means "twin," compare "Gemini"; Robert Recorde, *Whetstone of Witte* (London, 1557), quoted in Florian Cajori, *A History of Mathematical Notations. Vol. 1: Notations in Elementary Mathematics* [1928] (New York, 1993), p. 165.

325. Cf. Chapter 3.

Michel Foucault's *Archaeology of Knowledge* aimed at localising *dispositifs*, objective symbolic structures whose effectiveness was apparent through a multitude of concrete phenomena like regulations, truths, architectures, etc. that they generated, comparable in principle to the workings of the Hegelian *Weltgeist* (world spirit).[326] The more the humans that constitute the *Weltgeist* achieve their aims through instruments, investigate reality with apparatuses, and change it through the material associations of technology, the more the history of ideas is forced into matter. An archaeology of algorithmic artefacts endeavours to reconstruct the theoretical currents from the objective technical forms that generated them and were generated by them.

It is a pity that the Polish bomba never found its way onto Italian soil. Some ingenious owner of a latteria might have discovered that the construction, with its planetary gear and six mixers, was even better suited for whipping cream than a motor from a smashed-up motorbike, and might have adopted it, thereby making the strange legends entwined around the device come full circle.

9. Postscriptum

One of the most enlightening interpretations of Odradek says that its name is an incomplete anagram of the Greek word "dodeka(ed)r(on)," one of the five perfectly regular Platonic solids.[327] This proposal can be slightly extended. Apart from the metaphysical speculations of Kepler, this figure does not play any practical role in the sciences, for example, in crystallography. Yet almost one hundred so-called "Roman dodecahedra" from the second to the fourth century AD made of brass have been found within the borders of the former Roman Empire, from England to Hungary and the east of Italy, but most of them in Germany and France (cf. Figure 17): "They are hollow with circular holes of differing sizes in the faces and with knobs at their vertices."[328] The first specimens were already described in 1896. The function of these objects is completely unknown to the present day; none of the various hypotheses have been confirmed, and most have been disproved.[329]

Fig. 17: Roman dodecahedron.

326. Michel Foucault, *The Archaeology of Knowledge* [1969], trans. A.M. Sheridan Smith (London, 1972).

327. Cf. Jean-Claude Milner, Odradek, la bobine de scandale. *Elucidation* 10 (2004): 93–96.

328. Benno Artmann, Roman dodecahedra. *The Mathematical Intelligencer* 15. 2 (1993): 52–53.

329. The objects in Figure 10 are a Geissler tube magnetic motor, France, 1870, an electromagnetically contracting helix, Daniel Davis, 1848, and an unusual electric motor, England, 1860. John D. Jenkins generously provided the high-resolution images.

Resurrecting the *Bomba Kryptologiczna*:
A Reconstruction of the Polish Crypto Device

Shortly before the Second World War, the Polish mathematicians Marian Rejewski, Henryk Zygalski and Jerzy Różycki of the German section BS-4 of the *Biuro Szyfrów* devised a semi-automatic device to break the German "Enigma." Although this is by now an established fact,[330] the exact cryptanalytic method employed remains remarkably obscure, and there is no detailed description in Rejewski's accounts. The aim of this chapter is to shed light on this important early stage in the attack on the German encryption device by simulating the Polish artefact in software and trying to determine a procedure simple enough to solve the rotor order, ring setting and Steckers within the reported two hours.[331]

From September 1938 to May 1940, Enigma was employed in the following way: For each day, the operator on the sending side would locate the order of the three rotors, the five to eight pairs of letters to be permuted by the plugboard, and the so-called *Ringstellung* on a sheet listing the settings for the month. (The circumferential alphabet could be rotated with respect to the core of the wheels and its inner wiring. When the right ring was advanced one step, the permutation that had been at indicators AAA was now found at AAB, etc.) He "randomly" selected a *Grundstellung* (basic setting) on his own that was transmitted two times in clear, followed by the doubled message key encrypted with it and the telegram enciphered with the latter.[332] The result was communicated acoustically in Morse code on short wave radio links and could be intercepted by the listening stations of the Poles, the British, and other European countries.

---

330. This has not always been the case, as the heated discussion almost 40 years after the end of the war shows, cf. Marian Rejewski, Remarks on Appendix 1 to British Intelligence in the Second World War by F.H. Hinsley. *Cryptologia* 6. 1 (1982): 75–83. The Polish original, Uwagi do Appendix 1: The Polish, French and British Contributions to the breaking of the Enigma książki: British Intelligence in the Second World War prof. F.H. Hinsley'a, can be found at http://www.spybooks.pl/en/enigma.html.

331. M. Rejewski, The mathematical solution of the Enigma cipher. Appendix E, in: Władysław Kozaczuk, *Enigma: How the German Machine Cipher Was Broken, and How It Was Read by the Allies in World War Two* (Frederick, MD, 1984), pp. 272–291, p. 290.

332. For a detailed description of the machine and the procedures employed at the time, see Kozaczuk, *Enigma*, p. 247ff.

The Machine

In their attack, the Polish codebreakers relied on the phenomenon that in a number of intercepts, one letter of the message key, which was doubled to protect it against transmission errors, was encrypted to the same character. This resulted in groups like **W**AV**W**HA, in which the first symbol was enciphered as W on the first and the fourth position. These doublings were called "females" by the Poles and also occurred at the second and fifth or third and sixth letter.[333] According to Kozaczuk, the equivalent Polish term "samiczka, Plur. samiczki" resulted from a diminutive of the word "te same" (the same). It was later adopted by the British, who ignored the meaning of the term.[334] Females could be employed to deduce the ring setting of the wheels, to recover the plugs and finally the message key, allowing all the dispatches of a day to be read, but 105,456 rotor positions had to be searched for a specific pattern. If manual testing of an indicator would have taken a minute, the time needed for the whole operation would have amounted to more than two months, and by then, the content of the messages would have been strategically worthless.

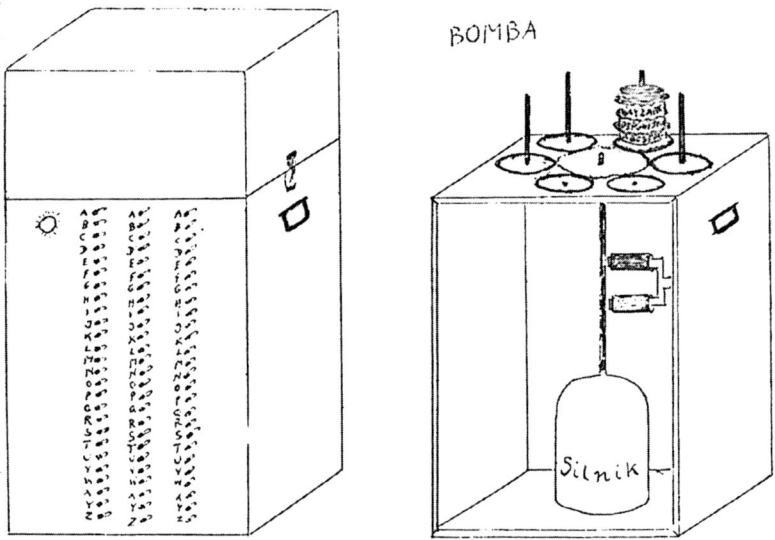

Fig. 1: Outside and interior of Polish "bomba," drawing by Rejewski, 1978.[335]

333. The modern term is "fixpoint," cf. Friedrich L. Bauer, *Decrypted Secrets. Methods and Maxims of Cryptology* (Berlin, 2000), p. 396.

334. Kozaczuk, *Enigma*, p. 63.

335. Brian Johnson, *The Secret War* (London, 1978), p. 316.

Consequently, the Poles mechanised the task. The *bomba kryptologiczna* ("cryptologic bomb") consisted of three pairs of Enigma rotor sets driven by an electric motor via a planetary gear. Six "bomby" (Polish plural of "bomba") were quickly built by the AVA Radio Manufacturing Company ("Wytwórnia Radiotechniczna AVA") in Warsaw in November 1938, one for each of the possible wheel orders. The firm, which at one point employed more than 200 people, had previously built the Polish rotor encryption device Lacida, Enigma doubles and another cryptological apparatus, the Cyclometer. Apparently, at least some copies of the first artefact have survived.[336]

The offsets of the simultaneities were set up on the hardware of the machine. If three dispatches beginning GKD WAVWHA, JOT IWABWN and MDO OTWYZW had been received, the first three letters of each being the Grundstellung in clear and the last six the encrypted doubled message key containing females on positions 1–4, 2–5 and 3–6, it was "programmed" corresponding to Figure 2.[337]

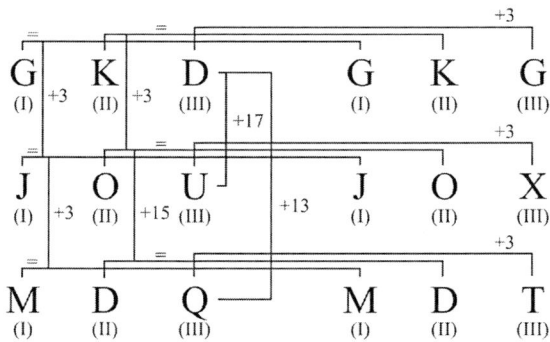

Fig. 2: Bomba setup for breaking females of indicators GKD, JOT and MDO.

336. Kozaczuk, *Enigma*, pp. 211, 134, 25, 263.

337. The example is taken from the earliest of Rejewski's accounts, M. Rejewski, Enigma 1930–1940. Metoda i historia rozwiązania niemieckiego szyfru maszynowego (w zarysie) ("Enigma 1930–1940. Method and history of the solution of the German cipher machine (in outline)," 32 page typescript, in Polish, ~1940; online: http://www.spybooks.pl/en/enigma.html), p. 29. The material on this website originates from the private files of Col. Władysław Kozaczuk (Rafal Brzeski, personal communication, 3 April 2007). The document was prepared in France to be included into an internal report by the head of the Cipher Bureau, Col. Gwido Langer, dated 12 May 1940. The same indicators are repeated in M. Rejewski, Matematyczne podstawy konstrukcji bomb kryptologicznych oraz uwagi o ich wykorzystaniu w Polsce i w Wielkiej Brytanii. Informacje o pierwszym komputerze na świecie ("The mathematical foundation of the construction of the cryptologic bombs and notes on their employment in Poland and England. Informations on the first computer of the world," 2 page typescript, in Polish, 1979, probably written in preparation for Kozaczuk's book; online: http://www.spybooks. pl/en/enigma.html), p. 1, and Tadeusz Lisicki, Appendix, in: Józef Gárlinski, *Intercept. The Enigma War* (London, 1979), pp. 192–204, p. 203. Most discussions of the bomba quote the example RTJ-DQX-HPL, given much later in an article published posthumously, M. Rejewski, Jak matematycy polscy rozszyfrowali Enigmę ("How Polish mathematicians broke the Enigma cipher," in Polish). *Wiadomości Matematyczne* 23. 1 (1980): 1–28. English translation: How the Polish mathematicians broke Enigma. Appendix D, in: Kozaczuk, *Enigma*, pp. 246–271, p. 266. Cf. for example Bauer, *Decrypted Secrets*, p. 395f. The authenticity of these examples will be investigated below.

Different witnesses have given various explanations for the name of the device, "bomba kryptologiczna," but none are easy to believe. Rejewski wrote they had designated it thus "for lack of a better idea."[338] After escaping from occupied France through Spain, Portugal and Gibraltar, the two remaining cryptanalysts worked for a Polish signal battalion stationed in Stanmore, England, from August 1943 on. (Jerzy Różycki had perished in the sinking of a passenger ship when returning from an outpost in Algiers to France on 9 January 1942.) Its commander, Tadeusz Lisicki, came up with the most popular legend; the mathematicians had come up with the idea of the device while eating a popular ice-cream dessert called "bomba."[339] The third explanation is found in an internal US American military report written four years after the bomby had been destroyed, on 11 October 1943: "This term ['bombe'] was used by the Poles and has its origin in the fact that on their device when the correct position was reached a weight was dropped to give the indication."[340]

The Procedure

Usually, the following account of the procedure executed on the bomba is given: The operator set the first pair of Enigmas to the indicators that had produced the coincidence at the first and fourth position of the doubled key, GKD and GKD+3 = GKG. The offset of three reflected the fact that in females, the same letter was produced three steps apart. He then turned the second pair to JOU, the indicator at which the second simultaneity had occurred, JOT, augmented by one, because it had happened one step later than the first (at symbols two and five). Again, he offset the second machine of the couple by three relative to the first, JOX. The same consideration led to the indicators MDQ (MDO+2) and MDT (MDQ+3) for the third pair of Enigmas. Current was switched on at the contact corresponding to the repeated letter of the females (in our example, W), and the apparatus automatically stepped through all possible wheel positions of one rotor order, until each couple output two identical characters and reproduced the three-fold simultaneity.[341] In this case, the machine stopped lighting a lamp, which permitted the operator to note the indicators.[342] Revolving the rotors through all possible 17,576 positions on six bomby in parallel each working on one wheel order took about 100–120 minutes. That allows the estimate that it was testing

338. Rejewski, How the Polish mathematicians, p. 267.

339. Letter by Tadeusz Lisicki, 30 August 1982, quoted in Kozaczuk, *Enigma*, p. 63, fn. 1.

340. Anonymous (Commander Howard T. Engstrom?), Note on early bombe history, ed. Frode Weierud (typescript, 11 October 1943, National Archives and Records Administration, RG 457, NSA Historical Collection, Nr. 1736, CBLH17, Box 705; online: http://cryptocellar.org/Enigma/BombeHistNote.pdf), p. 1.

341. The notches that moved the neighbouring wheel forward had all been set to Z to allow regular turning of the rotors. Cf. Heinz Ulbricht, *Chiffriermaschine Enigma. Trügerische Sicherheit. Ein Beitrag zur Geschichte der Nachrichtendienste* (Ph.D. diss., Technical University Braunschweig, 2005; online: http://www.digibib. tu-bs.de/?docid=00001705), p. 101.

342. Kozaczuk, *Enigma*, p. 53.

two to three settings per second, at a speed between 5.6 and 6.8 rpm (revolutions per minute).[343] From the indicators and rotor order found, the ring setting, plugs and message key could be deduced, allowing all the dispatches of a day to be read. At a conference near Warsaw in late July 1939, shortly before their country was invaded by German troops, the Poles handed two Enigma replicas, technical drawings of the Cyclometer and the Bomba over to the British and French.[344] All bomby were destroyed in September 1939, before the cryptologists were forced to flee over Romania, Serbia, Croatia and Italy to Paris.[345]

It is an irritating fact that the procedure published in Rejewski's accounts and subsequently in many of the books on the topic does not effectively solve Enigma cryptograms when tried out on an emulator. Most descriptions assume that one of the indicator settings the machine had halted at would translate the enciphered six letters into "something of the form XYZXYZ," the doubled message key in clear.[346] In 2005, Heinz Ulbricht, who served in the German Air Force during World War II, submitted his detailed and methodically novel Ph.D. dissertation at Technical University Braunschweig, in which he simulated all methods used by the Poles and British with computer programs.[347] In his account of the bomba, he omitted the question how to convert the rotor positions found into ring settings by directly turning the circumferential alphabets in his software, an operation technically impossible on the original machine. But even then, the resulting indicators do not easily allow for the reconstruction of the message key, because around half of the symbols, three of six, were steckered on the plug board at the time.[348] After five pages of complicated argumentation and experiment, Ulbricht arrived at a solution for the cryptogram. It seems improbable that the method described by him allowed the Poles to read Enigma traffic on a regular basis. When his example was set up on the author's bomba emulator, it did not stop at all in the wheel order it was based on. The reason was that the last of his basic settings, UQR, produced a turnover of the middle and left rotor while encrypting the doubled message key, since the notch in the middle wheel, number 1, is at Q. It is known that two conditions are necessary for the procedure to work:

343. Cf. Rejewski, The mathematical solution, p. 290. The speed of its successor, the Turing bombe, was about 10 times higher, testing 26.6 indicators per second, at a speed of 61.5 rpm: "Our machine [the Turing bombe] will complete a full unsuccessful run in 11 minutes." (John Harper, personal communication, 31 March 2008)

344. Kozaczuk, *Enigma*, p. 59. In the National Archives, London, ref. HW 25/9, a document is located named "ZYKLOMETER schematic," probably being the only remaining or at least the only declassified item of the exchange, annotated in perfect German. The location of the two Enigma doubles is unknown. The replica in Sikorski Institute, London, was manufactured at a later date by a French company, as proven by the word "Controle" printed on the top right corner. Cf. Kozaczuk, *Enigma*, p. 178, caption of photo on the right side. The history of cryptology would be furthered if these items were finally, 75 years after they were handed over, declassified.

345. Kozaczuk, *Enigma*, p. 77, fn. 5, and Rejewski, Remarks on Appendix 1, p. 81.

346. B. Jack Copeland, Enigma, in: *The Essential Turing: Seminal Writings in Computing, Logic, Philosophy, Artificial Intelligence, and Artificial Life*, ed. B.J. Copeland (Oxford, 2004), pp. 217–264, p. 244.

347. The simulative work has added many valuable details on the concrete operations in codebreaking Enigma. It is typical for the exact history of algorithmic artefacts that it is almost impossible to understand the dissertation without performing the procedures and consequently, reconstructing the machines, at least in software.

348. Ulbricht, *Chiffriermaschine Enigma*, pp. 100–106.

1) The repeating character in the female must not be steckered. The fulfilment of this condition is due to chance. If 5 to 8 pairs of letters are exchanged, as was the case in November 1938 when the bomba was built, it will be met half the time on average.[349]

2) Only the right-hand wheel can move during encryption of the six signs of the doubled message key.[350] Since the positions of the notches on the rotors were known at the time, this was easy to arrange. For wheel order 312, the last letter of the basic setting has to be smaller than Z and greater than E, because drum 2 steps its neighbour forward at E, and the middle indicator should not be Q, its turnover position.

Reconstructing the Polish Routine

Fortunately, Rejewski provided a theoretical hint at the method of solution in his discussion of the bomba:

> "Let us assume, for a moment, that permutation S [the exchange of letters on the plugboard] is identical. If, as well, there were no setting of rings, and we also knew the sequence of the drums on the axle, it would suffice to set the drums to position RTJ [the basic key of the first female], and a depression of the key 'w' would cause one and the same lamp to light within the interval of three strokes. The same would happen in position HPN and position DQY [the basic keys of the other two females incremented by their relative offsets] of the drums. The setting of the rings causes the positions of the drums at which this will occur to be unknown to us. However, the differences in the positions will be preserved, and are therefore known to us."[351]

After programming the emulator of the bomba, numerous experiments attempting to generate message keys and trying to break them were unsuccessful, because the indicators at which it came to a halt were not easily interpreted.[352] Trying to simplify the process, females for ring setting AAA were produced and run on the virtual machine.[353] It stopped at the starting position, and proved Rejewski's above statement in a practical way: If the circumferential alphabets are not rotated, the solution is produced immediately. For females of ring setting AAB the machine halted at the very last position, and for AAZ at the second, always relative to the Grundstellung of the 1–4 repetition.

---

349. Ulbricht, *Chiffriermaschine Enigma*, p. 101.

350. Copeland, Enigma, p. 241.

351. Rejewski, How the Polish mathematicians, p. 266.

352. The program can be found on the author's website, http://www.alpha60.de/research/. In parallel to the historical developments, the Enigma simulator Andy Carlson has provided at http://www.mtholyoke.edu/~a-durfee/cryptology/enigma_j.html has been reverse-engineered to assemble a bomba out of six German encryption machines.

353. The other settings were: wheel order 213, plugs AM, CQ, DF, EY, HL, JX, OZ. All cryptograms in this paper are based on reflector ("Umkehrwalze") B, which was the one employed from autumn 1937 on (Rejewski, How the Polish mathematicians, p. 264).

Ring setting	Females		Message key in clear	Start position (a)	Stop position (b)	c = stop − start	AAA − c
AAA	HDW	**B**JZ**B**SR	HRF	HDW	HDW	ZZZ	AAA
	OOC	**BB**LQ**B**Z	NYT				
	TWB	LM**B**VE**B**	RQI				
AAB	FUE	**B**JI**B**US	NZL	FUE	FUD	ZZY	AAB
	PTB	A**B**ZI**B**D	WZX				
	TZD	AF**B**FV**B**	HHI				
AAZ	HGJ	**B**U**BB**SI	IHC	HGJ	HGK	ZZA	AAZ
	WAA	S**B**CF**B**Y	MYU				
	OCX	AE**B**GD**B**	NCY				
AZA	YSX	**B**ED**B**PA	PRK	YSX	YTX	ZAZ	AZA
	QLC	Y**B**IQ**B**R	PHB				
	FHF	FV**B**II**B**	DTM				
ZAA	ZKE	**B**PA**B**AE	NCL	ZKE	AKE	AZZ	ZAA
	ALY	**BB**NZ**BB**	TKZ				
	CWI	RW**B**QP**B**	NBF				
XYZ	IKW	**B**DT**B**OX	RNU	IKW	LMX	CBA	XYZ
	FPZ	W**B**WS**B**G	LTC				
	HXB	QW**B**AG**B**	JZE				

Fig. 3: Bomba stops for different ring settings and their interpretation.

Experimental evidence suggested it was not the indicators at which the machine stopped that were of importance, but the offset from the starting position at which that occurred. The ring setting could then be directly derived by subtracting this value from AAA.

An Authentic Message

To illustrate the full decoding process, the authentic message provided by David Kahn and first published by Cipher A. Deavours and Louis Kruh will now be broken using Polish methods and the bomba emulator.[354] It was sent by Generaloberst Walther von Brauchitsch to Heeresgruppenkommando 2 (later Army Group C) at Frankfurt am Main on 21 September 1938.

---

354. Cipher A. Deavours and Louis Kruh, The Turing Bombe: Was it enough? *Cryptologia* 14. 4 (1990): 331–349, p. 342. Frode Weierud has published important corrections to this article on his website, http://cryptocellar.org/Enigma/tbombe.html. Frank Carter from Bletchley Park has provided a similar account of the Polish methods, which I was not aware of when investigating the bomba, cf. Frank Carter, *The First Breaking of Enigma. Some of the Pioneering Techniques Developed by the Polish Cipher Bureau* (Bletchley Park Report No. 10, Milton Keynes, July 1999). There is only a minor difference in the calculation of ring settings, his "null" position being ZZZ and mine, AAA.

The three parts of the encrypted message read:

```
Fernschreiben H.F.M.No. 563
+ HRKM 13617 1807 =
AN HEERESGRUPPENKOMMANDO 2=
2109 -1750 - 3 TLE - FRX FRX - 1TL -172=
HCALN UQKRQ AXPWT WUQTZ KFXZO MJFOY RHYZW VBXYS IWMMV WBLEB
DMWUW BTVHM RFLKS DCCEX IYPAH RMPZI OVBBR VLNHZ UPOSY EIPWJ TUGYO
SLAOX RHKVC HQOSV DTRBP DJEUK SBBXH TYGVH GFICA CVGUV OQFAQ WBKXZ
JSQJF ZPEVJ RO -


2TL - 166 - ZZWTV SYBDO YDTEC DMVWQ KWJPZ OCZJW XOFWP XWGAR
KLRLX TOFCD SZHEV INQWI NRMBS QPTCK LKCQR MTYVG UQODM EIEUT VSQFI
MWORP RPLHG XKMCM PASOM YRORP CVICA HUEAF BZNVR VZWXX MTWOE
GIEBS ZZQIU JAPGN FJXDK I -


3 TL - 176 - DHHAO FWQQM EIHBF BMHTT YFBHK YYXJK IXKDF RTSHB
HLUEJ MFLAC ZRJDL CJZVK HFBYL GFSEW NRSGS KHLFW JKLLZ TFMWD QDQQV
JUTJS VPRDE MUVPM BPBXX USOPG IVHFC ISGPY IYKST VQUIO CAVCW AKEQQ
EFRVM XSLQC FPFTF SPIIU ENLUW O

=  1 ABT GEN ST D H NR. 2050/38 G KDOS +
```

The Grundstellung chosen by the operator was FRX, and the crossed-out first six letters of each segment represent the encrypted message keys, which constitutes the following indicators:[355]

1. FRX HCALNU
2. FRX ZZWTVS
3. FRX DHHAOF.

Since additional keys from the day in question have not been preserved, the author has generated these using the wheel order, ring setting and Stecker connections of the message.[356] On that day, the Poles could have received the following three females,

4. BOP **A**DD**A**KS
5. KFY **I**AQ**H**AU
6. IDB PN**A**OU**A**,

---

355. The crossed-out third group in each part was the identification group ("Kenngruppe") designating the cipher net the dispatch was adressed to. Cf. Oberkommando der Wehrmacht (OKW), Schlüsselanleitung zur Schlüsselmaschine Enigma (H.Dv.g. 14, 13 January 1940; online: http://www.ilord.com/enigma-manual1940-german.pdf), p. 6f.

356. Also the indicator generator can be found on the author's website, http://alpha60.de/research/.

and six further message keys,

```
7.    AAA  QZMOMS
8.    ABC  RQBKQR
9.    OKW  OQEUMA
10.   REX  VSERNC
11.   NAX  DJWLOO
12.   KFZ  XOHYST.
```

The Enigma doubles in each of the six bomby (each testing one of the six possible wheel orders) are set to indicators

```
I.     BOP  -  II.    BOS
III.   KFZ  -  IV.    KFC
V.     IDD  -  VI.    IDG
```

and are rotated through all positions while applying current to the letter repeated in the females, A. The machines stop at the following indicators:

Wheel order	Bomba stop and output	Ring setting (AAA − stop + BOP)
123	**OVO** OVR XMY XMB VKC VKF → RREEZZ	**NTB**
132	**MQU** MQX VHE VHH TFI TFL → VVCCFF	**PYV**
	UFH UFK DWR DWU BUV BUY → DDXXZZ	**HJI**
213	**CSM** CSP LJW LJZ JHA JHD → NNJJII	**ZWD**
231	**WVQ** WVT FMA FMD DKE DKH → CCTTJJ	**FTZ**
312	**GJE** GJH PAO PAR NYS NYV → CCFFSS	**VFL**
321	℅	℅

Fig. 4: Bomba results and corresponding ring settings.

In the first column, the rotor order is found, and in the second, the halting positions of the six Enigmas in the bomba, followed after the arrow by the three couples of identical letters they produce. To calculate the ring settings in the last column from the stop positions, the table from Figure 5 was employed. For the first halt OVO with start position BOP, the first letter of the latter indicator, B (2), needs to be subtracted from O (15), which equals 13. Looking up 13 in the third row of the table, N is obtained. If the difference is negative, the fourth row can be used. Correspondingly, V (22) − O (15) = 7 (T) and O (15) − P (16) = −1 (B) is calculated, which results in ring setting NTB. The Beaufort table Deavours and

Kruh provided in their article on the Turing bombe represents an equivalent of this procedure.[357]

1	2	3	4	5	6	7	8	9	10	11	12	13	14	15
A	B	C	D	E	F	G	H	I	J	K	L	M	N	O
0	25	24	23	22	21	20	19	18	17	16	15	14	13	12
0	-1	-2	-3	-4	-5	-6	-7	-8	-9	-10	-11	-12	-13	-14

16	17	18	19	20	21	22	23	24	25	26
P	Q	R	S	T	U	V	W	X	Y	Z
11	10	9	8	7	6	5	4	3	2	1
-15	-16	-17	-18	-19	-20	-21	-22	-23	-24	-25

Fig. 5: Table for the conversion of indicators to ring settings.

Now the message keys of the three segments of the cryptogram (FRX HCALNU, FRX ZZWTVS, FRX DHHAOF) will be deciphered with the different wheel orders and ring settings obtained.[358] By counting how many females arise, it can be determined which of the above alternatives is the most probable one:

	W123, NTB=14,20,2	W132, PYV=16,25,22	W132, HJI=8,10,9	W213, ZWD=26,23,4
FRX HCA LNU	BOV KQO	ALK XRZ	KLY SJS	AGU MGI
FRX ZZW TVS	AJO RMR	IDR HLR	YFJ RIU	LDF YWZ
FRX DHH AOF	JXU BFW	WFL CXK	MIG KUB	BIN BDN
	0	1	0	3

W231, FTZ=6,20,26	W312, VFL=22,6,12
NWK FUY	PLG NRK
PSB OCM	NBC RFW
JES YSX	FPR DKH
0	0

Fig. 6: Experimental decipherment of message keys with solutions obtained.

---

357. Deavours and Kruh, Turing Bombe, p. 335.

358. Here and in the following I relied on Dirk Rijmenants' excellent Enigma simulator (http://users.telenet. be/d.rijmenants/en/enigmasim.htm) for manual ciphering.

The fourth ring setting and rotor sequence bring forth significantly more letter doublings than the others, which suggests they are the ones sought-after. If needed, more indicators can be included into the test.

Deriving Steckers

Since it is known that the cleartext of the key was "something of the form XY-ZXYZ," the Stecker connections can now be derived by decrypting the indicators received with the ring setting and wheel order that has just been determined. For every sign that does not repeat on positions 1–4, 2–5 and 3–6, there exist two cross-pluggings that establish identity. The input letter that produced the difference needs to be connected to another one that converts to the desired output symbol, equal to the one at the other place. Without plugs, **H**CALNU decrypts to **A**GUMGI in basic key FRX. Taking positions 1–4 as example, either H needs to be cabled to U, which produces M in setting FRX (the indicator the machine is at when permuting the first sign), and then **H**CALNU deciphers to **M**GUMGI; or L needs to be connected to B, which converts to A in setting FRA (the indicator at the fourth letter), then **H**CALNU decrypts to **A**GU**A**GI. To decide which of the two Stecker connections to plug, we derive the alternatives for all differing symbols in the keys that have been received. Due to the reciprocity of Enigma, this can be achieved by simply typing the target letter at corresponding indicators, i.e., by setting the machine to FRX and typing **M**XX**A**XX, resulting in **U**XX**B**XX for the above example.

Basic setting and encrypted message key	Decrypted message key	Plug alternatives	Resulting plug
1.FRX HCA LNU	AGU MGI	1-4: H-U v L-B	
		3-6: A-E v **U-I**	
2.FRX ZZW TVS	LDF YWZ	1-4: Z-E v T-M	
		2-5: Z-R v V-O	
		3-6: W-R v S-N	
3.FRX DHH AOF	**BIN BDN**	2-5: H-Z v O-J	
4.BOP ADD AKS	**NWO NWO**	%	
5.KFY IAQ HAU	FJW VJO	1-4: **I-U** v H-I	**I-U**

Fig. 7: Deduction of plug alternatives from decryption of message keys.

We find that plug `I-U` should be set, because it appears twice. The Stecker alternatives converge. Repeating the procedure with the letters exchanged reveals one more junction:[359]

Basic setting and encrypted message key	Decrypted message key	Plug alternatives	Resulting plug
1.FRX HCA LNU	**AGI** MGI	1-4: H-I v L-B	
2.FRX ZZW TVS	LDF YWZ	1-4: Z-E v T-M	
		2-5: Z-R v V-O	
		3-6: W-R v S-N	
3.FRX DHH AOF	**BUN BDN**	2-5: H-Z v **O-J**	
5.KFY IAQ HAU	**VJW VJW**	%	
6.IDB PNA OUA	HZU S**ZU**	1-4: P-K v O-X	
7.AAA QZM OMS	PKY YCE	1-4: Q-E v **O-J**	**J-O**

Fig. 8: Further deduction of plugs from decryption of message keys.

We continue the process with Steckers `I-U` and `J-O` plugged, realising each connection that appears twice.

Basic setting and encrypted message key	Decrypted message key	Plug alternatives	Resulting plug
1.FRX HCA LNU	**AGI MGI**	1-4: H-I v L-B	
2.FRX ZZW TVS	LDF YWZ	1-4: Z-E v T-M	
		2-5: Z-R v V-J	
		3-6: **W-R** v S-N	
3.FRX DHH AOF	**BUN BUN**	%	
6.IDB PNA OUA	HZU **Q**ZU	1-4: P-X v O-X	
7.AAA QZM OMS	**P**KY **P**CE	2-5: Z-L v M-N	
		3-6: M-V v S-I	
8.ABC RQB KQR	HSS ASV	1-4: **R-W** v K-M	**R-W**

---

359. Here and in the following, message keys that already decrypt to a repeated identical sequence have been omitted.

Cycle IV.

1.FRX HCA LNU	**AGI** M**GI**	1-4: H-I v L-B	
2.FRX ZZW TVS	LD**Z** YR**Z**	1-4: Z-E v T-M	
		2-5: Z-W v V-J	
6.IDB PNA OUA	H**ZU** Q**ZU**	1-4: P-X v O-X	
7.AAA QZM OMS	**P**KY **P**CE	2-5: Z-L v M-N	
		3-6: **M-V** v S-I	
8.ABC RQB KQR	**AS**S **AS**D	3-6: B-L v R-E	
9.OKW OQE UMA	**L**RL **LL**D	2-5: Q-G v **M-V**	**M-V**

Cycle V.

1.FRX HCA LNU	**AGI** V**GI**	1-4: H-I v L-B	
2.FRX ZZW TVS	LD**Z** YL**Z**	1-4: **Z-E** v T-V	
		2-5: **Z-E** v V-J	**E-Z**

Cycle VI.

1.FRX HCA LNU	**AGI** V**GI**	1-4: H-I v **L-B**	
2.FRX ZZW TVS	**YLE YLE**	%	
6.IDB PNA OUA	H**EU** Q**EU**	1-4: P-X v O-X	
7.AAA QZM OMS	**PYZ PYZ**	%	
8.ABC RQB KQR	**AS**S **AS**D	3-6: **B-L** v R-Z	**B-L**

Cycle VII.

1.FRX HCA LNU	**AGI AGI**	%	
6.IDB PNA OUA	H**EU** Q**EU**	1-4: **P-X** v O-X	
8.ABC RQB KQR	**ASD ASD**	%	
12.KFZ XOH YST	S**AM** I**AM**	1-4: **X-P** v Y-Z	**P-X**

Cycle VIII.			
1.FRX HCA LNU	**AGI AGI**	%	
2.FRX ZZW TVS	**YBE YBE**	%	
3.FRX DHH AOF	**LUN LUN**	%	
4.BOP ADD AKS	**NRJ NRJ**	%	
5.KFY IAQ HAU	**MOR MOR**	%	
6.IDB PNA OUA	**QEU QEU**	%	
7.AAA QZM OMS	**XYZ XYZ**	%	
8.ABC RQB KQR	**ASD ASD**	%	
9.OKW OQE UMA	**BRD BRD**	%	
10.REX VSE RNC	**DUM DUM**	%	
11.NAX DJW LOO	**SPD SPD**	%	
12.KFZ XOH YST	**IAM IAM**	%	

Fig. 9: Further deduction of plugs.

At this point, all sequences decrypt to doubled indicators in cleartext, all Stecker connections are found, and in combination with the rotor sequence and the ring settings already obtained, all the dispatches of the day can be deciphered. The message keys of the three segments of the cryptogram, HCA LNU, ZZW TVS and DHH AOF, now decode to AGI AGI, YBE YBE and LUN LUN, and when the machine is set to AGI, the letters that follow the key,

```
QKRQ WUQTZ KFXZO MJFOY RHYZW VBXYS IWMMV WBLEB DMWUW BTVHM RFLKS
DCCEX IYPAH RMPZI OVBBR VLNHZ UPOSY EIPWJ TUGYO SLAOX RHKVC HQOSV
DTRBP DJEUK SBBXH TYGVH GFICA CVGUV OQFAQ WBKXZ JSQJF ZPEVJ RO,
```

are transformed to readable German:

```
AUFBE FEHLD ESOBE RSTEN BEFEH LSHAB ERSSI NDIMF ALLEX ZXZTX UNWAH
RSCHE INLIC HENXF RANZO ESISQ ENANG RIFFS DIEWE STBEF ESTIG
UNGEN JEDER ZAHLE NMAES SIGEN UEBER LEGEN HEITZ UMTRO TZZUH
ALTEN X
```

The same is true of the other parts of the message, which decrypt to

```
FUEHR UNGUN DTRUP PEMUE SSENV ONDIE SEREH RENPF LIQTD URQDR UNGEN
SEINX ABSXD EMGEM AESSB EHALT EIQMI RDIEE RMAEQ TIGUN GZURP UFGAB
EDERB EFEST IGUNG ENODE RAUQV ONTEI LENAU SDRUE CKLIQ
```

126

and

```
PERSO ENLIQ VORXA BSXAE NDERU NGDER ANWEI SUNGX OKHXG ENXST XDXHX
ERSTE ABTXN RXDRE IDREI ZWOEI NSXDR EIAQT GXKDO SXVOM JULIE INSNE
UNDRE IAQTB LEIBT VORBE HALTE NXDER OBERB EFEHL SHABE RDESH EERES
```

respectively, and this results in the following text in clear:

```
Fernschreiben H.F.M.No. 563
+ HRKM 13617 1807 -
AN HEERESGRUPPENKOMMANDO 2=
2109 -1750
```

Auf Befehl des Obersten Befehlshabers sind im Falle z. Zt. unwahrscheinlichen französischen Angriffs die Westbefestigungen jeder zahlenmäßigen Überlegenheit zum Trotz zu halten. Führung und Truppe müssen von dieser Ehrenpflicht durchdrungen sein.

Demgemäß behalte ich mir die Ermächtigung zur Aufgabe der Befestigungen oder auch von Teilen ausdrücklich persönlich vor.

Änderung der Anweisung OKH/Gen/St/D/H Erste Abt. Nr. 3321/38 G KDos vom Juli 1938 bleibt vorbehalten. Der Oberbefehlshaber des Heeres.

```
= 1 ABT GEN ST D H NR. 2050/38 G KDOS +
```
[360]

The content and style of this message is rather unexpected as it was issued well before the war while the Siegfried line was still under construction; at this time the French Army did not represent an imminent threat to the Germans. However, on the very same day in Prague, delegates of England and France declared that if the Czechoslovaks continued to refuse ceding the Sudetenland to Germany, they would be responsible for a war in which their countries would not participate, forcing president Eduard Benesch to comply with Hitler's demands. The German High Command might have suspected this was a ruse.

The event is documented in the papers of Sir Eric Phipps, the British ambassador in Paris: "Phipps tel., 21 September 1938, 5.5 p.m. The Czechoslovak Government replied on 21 September that, 'under the pressure of urgent insistence culminating in (the) British communications of 21 September,'

---

360. Translation: The Commander-in-Chief orders as follows: In the case of French attacks on the Western fortifications, although unlikely at this moment, those fortifications must be held at all costs, even against numerically superior forces. Commanders and troops must be imbued with the honour of this duty. / Accordingly, I emphasise that I alone have the right to authorise the fortifications to be abandoned in whole or part. / I reserve the right to make changes to order OKH/Gen/St/D/H 1. Abt. Nr. 3321/38 G KDos of July 1938. The Commander-in-Chief of the Army.

they 'sadly accept(ed) the French & British proposals.'"[361] The Enigma message was transmitted 45 minutes later.

Examples by Rejewski

Using the same procedure, the earliest example by Rejewski mentioned above will be investigated to find out if it is authentic, that is, was produced on a replica Enigma. It provides the following females

1.   GKD **W**AV **W**HA
2.   JOT I**W**A B**W**N
3.   MDO OT**W** YZ**W**

and these further keys:

4.   KTL WOC DRB
5.   SVW KKM IYS
6.   EDC DSP LJC
7.   BWK **T**CA **T**OC
8.   GRA FDR YWD
9.   KJC B**S**W R**S**E
10.   SGF TEY ASR
11.   AGH MD**F** RL**F**
12.   JBR WLT SOQ.

The bomby are set up with the following offsets:

I.	GKD –	**II.**	GKG
III.	JOU –	**IV.**	JOX
V.	MDQ –	**VI.**	MDT

361. John Herman, *The Paris Embassy of Sir Eric Phipps: Anglo–French Relations and the Foreign Office, 1937–1939* (Sussex, 1998), p. 218, fn. 52.

The letter `W` is continuously input while rotating through all possible positions. They stop at the following wheel orders and indicators, resulting in ring settings:

Wheel order	Bomba stop and output	Ring setting (AAA – stop + GKD)
123	**QBG** QBJ TFX TFA WUT WUW → RRHHMM	**QJX**
132	**OUV** OUY RYM RYP UNI UNL → IIKKYY	**SQI**
213	%	
231	**RED** REG UIU UIX XXQ XXT → LLRRDD	**PGA**
312	**IDG** IDJ LHX LHA OWT OWW → AAAALL **IQS** IQV LUJ LUM OJF OJI → UUMMJJ **AUY** AUB DYP DYS GNL GNO → RRSSSS	**YHX** **YUL** **GQF**
321	%	

Fig. 10: Bomba results with corresponding ring settings for Rejewski example.

Decrypting the message keys suggests the first rotor sequence and ring setting should be further investigated:

	W123, QJX=17,10,24	W132, SQI=19,17,9	W231, PGA=16,7,1	W312, YHX=25,8,24
KTL WOC DRB	**LTS BAS**	SJB ATQ	PWO SPS	NLV CSH
SVW KKM IYS	SZH NUJ	GAT VBK	WPV CWH	PI**K** HW**K**
EDC DSP LJC	UVB JHV	GCK HHD	BPD IRO	KEA JNI
GRA FDR YWD	JLY LZB	MXQ VKE	EPO OKQ	ZPE DNR
SGF TEY ASR	**M**WU **M**XB	GJU WIW	ZCU XBA	KUP JXQ
JBR WLT SOQ	HGL OND	LFA EKO	CID OQI	REO ALE
BWK TCA TOC	WHZ HBP	HBX VHJ	WBU HML	JPJ LQL
KJC BSW RSE	SQI XNQ	GRG JQV	VTH SHR	NPI YDB
AGH MDF RLF	**E**IJ G**I**X	XRS BTQ	OAP JMT	UUS DR**S**
	3	0	0	2

W312, YUL=25,21,12	W312, GQF=7,17,6
GFR WLP	DUG THF
RCK HJH	BIF DOC
YDG EVF	CAD THM
SNU DLC	NVH ORN
BUV ITO	WIP RKO
TVX JND	VJV BGT
CLJ HPS	LNE XQL
QZG MRH	GZZ BWD
ZWJ YWH	YSU WCK
1	0

Fig. 11: Experimental decipherment of message keys with solutions obtained.

But the deduction of Stecker alternatives does not converge:

Basic setting and encrypted message key	Decrypted message key	Plug alternatives	Resulting plug
1.GKD **W**AV **W**HA	**R**IO **R**NB	2-5: A-V v H-B	
		3-6: V-F v A-H	
2.JOT I**W**A B**W**N	ZHS TAH	1-4: **I-A** v B-Y	
		2-5: W-F v W-G	
		3-6: A-N v N-M	
3.MDO OT**W** YZ**W**	WIM AVM	1-4: O-B v Y-D	
		2-5: T-B v Z-G	
7.BWK **T**CA **T**OC	WHZ HBP	1-4: T-P v T-U	
		2-5: C-Y v O-Z	
		3-6: **A-I** v C-J	**A-I**

Cycle II

1.GKD **W**AV **W**HA	**R**IO **R**NN	2-5: A-V v **H-R**	
		3-6: V-Z v A-H	
2.JOT I**W**A B**W**N	**T**HO **T**IH	2-5: W-F v W-G	
		3-6: A-N v N-F	
3.MDO OT**W** YZ**W**	WA**M** IV**M**	1-4: O-B v Y-D	
		2-5: T-B v Z-G	
7.BWK **T**CA **T**OC	WH**P** HB**P**	1-4: T-P v T-U	
		2-5: C-Y v O-Z	
9.KJC B**S**W R**S**E	SQA XNQ	1-4: B-M v R-Q	
		2-5: S-V v S-T	
		3-6: W-T v E-Z	
11.AGH MD**F** RL**F**	EA**J** GA**X**	1-4: M-X v **R-H**	**H-R**

Cycle III

1.GKD **W**AV **W**HA	**H**IO **H**IN	3-6: V-Z v A-R	
2.JOT I**W**A B**W**N	**T**RO **T**IR	2-5: W-F v W-G	
		3-6: A-N v N-F	
3.MDO OT**W** YZ**W**	WA**M** IV**M**	1-4: O-B v Y-D	
		2-5: T-B v Z-G	
7.BWK **T**CA **T**OC	WR**P** RB**P**	1-4: T-P v T-U	
		2-5: C-Y v O-Z	
9.KJC B**S**W R**S**E	SQA TNQ	1-4: B-U v R-Q	
		2-5: S-V v S-T	
		3-6: W-T v E-Z	
11.AGH MD**F** RL**F**	EA**J** EA**X**	3-6: F-I v F-I	
4.KTL WOC DRB	LTS BUS	1-4: W-A v D-R	
		2-5: O-G v R-J	?

Fig. 12: Deduction of plug alternatives from decryption of message keys.

Rejewski obviously created a fictitious example as a mere illustration of the decoding process.[362]

Hardware

Concerning the hardware of the device, among others, one inconsistency stands out: If the same letter was input into each of the three pairs of Enigmas that made up the machine, why are there three columns of switches on the outside, each representing a full alphabet? They would have permitted to enter different characters into each couple. Perhaps Rejewski planned a more general procedure that worked with all females, no matter which letter repeated, like the sheet method invented by Zygalski.[363] He might only have realised through practical experiments with the machine that the same sign needed to be entered, because it was less probable that a single symbol was changed on the plug board. And even if he was not sure that the procedure would work, the bomby could be tried in case only females with different letters had been received, with a slight chance that they would produce the right ring setting. Designing the artefact to be more general than necessary gave him the potential to counter future changes in procedure by the Germans without having to build new hardware. When Alan Turing devised the British version of the device in 1940, the "bombe," he successfully followed the same principle, anticipating that the doubling of the message keys on which all Polish methods relied would be given up shortly after.[364]

The following diagram tentatively depicts the circuitry within the bomba. For the sake of simplicity, it has been reduced to only four letters. The motor at the bottom of the picture drives the six Enigmas on top via a planetary gear (not shown). As in the encryption device, closing the manual switches mechanically disconnects the corresponding contact of the rotor from the relay below, preventing it to be activated immediately.

---

362. The author has also tried to solve the example with reflectors A and C, with the same result. Also the second set of message keys Rejewski provided has been investigated and found not to be authentic. The full report is located on the author's website, http://alpha60.de/research/.

363. Rejewski, The mathematical solution, p. 287ff. Tony Sale has published a working simulator of the sheets method online, cf. http://www.codesandciphers.org.uk/virtualbp/poles/poles.htm

364. Gordon Welchman, *The Hut Six Story. Breaking the Enigma Codes* (New York, 1982), p. 81.

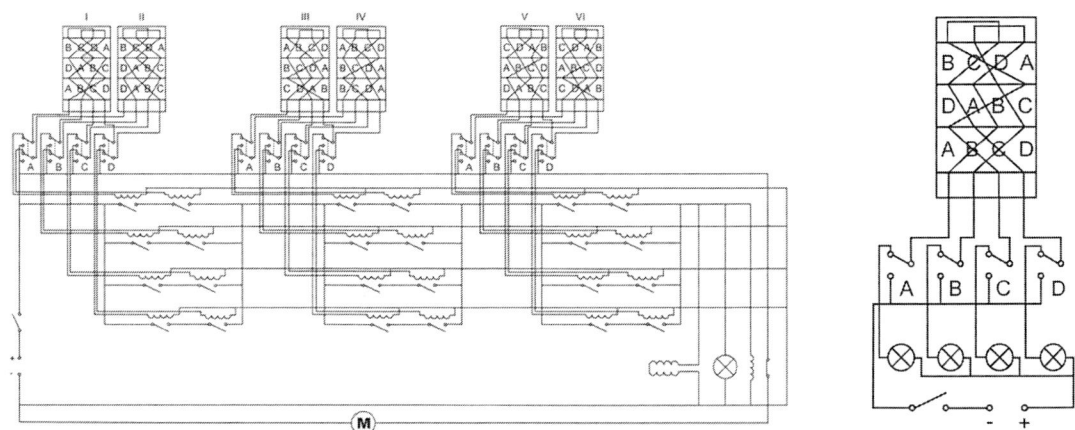

Fig. 13: Circuitry of a reduced version of the bomba in comparison with a reduced version of Enigma.

The couples of Enigmas in the simplified bomba are set up with an offset of three:

```
I.      BDA  -  II.     BDD
III.    ABC  -  IV.     ABB
V.      CAD  -  VI.     CAC.
```

This setup could be based on the following females (3-letter *Grundstellung* (basic setting) and 6 letters encrypted message keys):

```
1. BDA  BADBCA
2. ABB  ABDCBA
3. CAB  ADBCAB
```

The repeating letter in the females, B, is switched on in all three couples. In the first rotor position, the following relays close (cf. Figure 14):

```
I/II:   A-D
III/IV: D-C
V/VI:   C-D
```

133

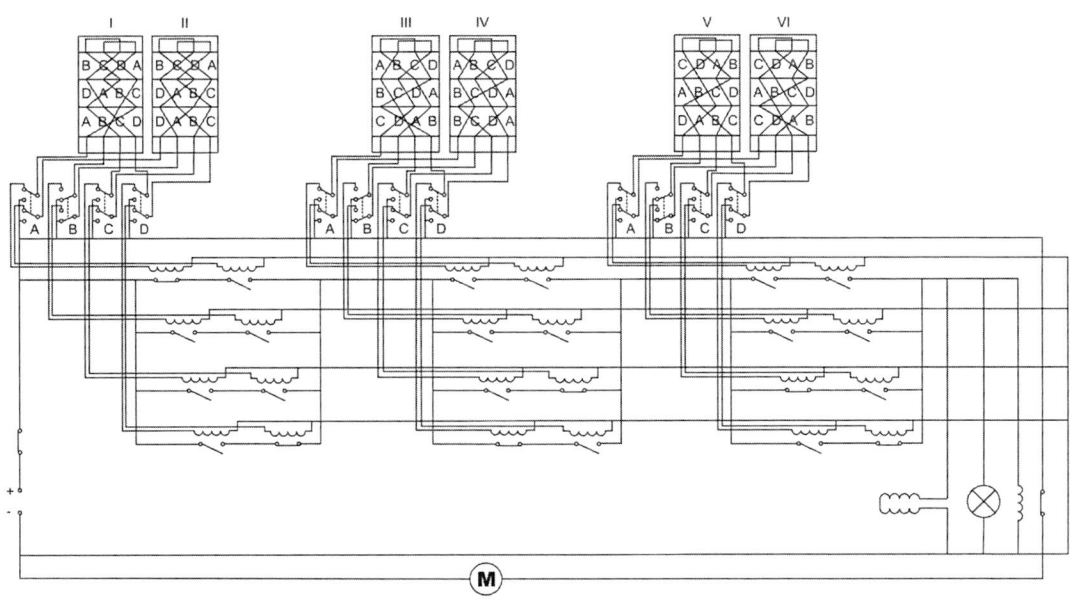

Fig. 14: Reduced version of bomba with current applied.

Nothing further happens. If, however, each of the three couples of Enigmas produces two identical letters, the circuit closes. This activates the horseshoe magnet, lights the lamp and opens the relay at the bottom right, disconnecting the motor from current. If the on/off switch at the bottom left is toggled, all relays in the middle are released and the system returns to its initial state.

The mysterious double solenoid visible in Rejewski's sketch (cf. Figure 1) close to the main axle shaft has been included into the circuitry, even though its function remains mysterious. It might have been part of a clutch mechanism to disengage the motor, or part of a system that produced the end result of the operation – the ring setting sought-after. A hardware equivalent of the above calculation would be a simple three-wheel counter stepping backwards from AAA each time the rotors of the apparatus moved forward. This complement would reinforce the device's similarity to a time bomb, which also counts down.

Scrambling T-R-U-T-H
Rotating Letters as a Form of Thought

"Know that the secrets of God and the objects of His science, the subtle realities and the dense realities, the things of above and the things from below, belong to two categories: there are numbers and there are letters. The secrets of the letters are in the numbers, and the epiphanies of the numbers are in the letters. The numbers are the realities of above, belonging to the spiritual entities. The letters belong to the circle of the material realities and to becoming."

Aḥmad Al-Būnī (d. 1225)[365]

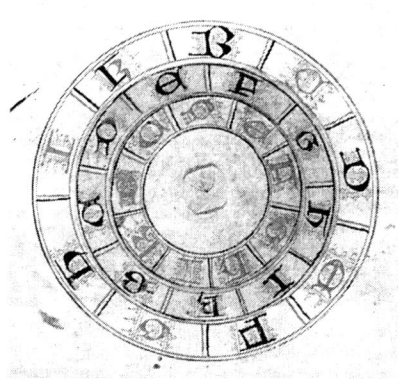

The works of the Majorcan philosopher and missionary Ramon Llull (1232–1316) are commonly and rightly regarded as foundational for the development of Western combinatorics and logic. Circular disks inscribed with the letters from B to K, which can be rotated in relation to each other, play a central role in his *Ars Inveniendi Veritatem*. A working model of the paper machine was included in several of his publications; a thread through the middle held the disks together.

Fig. 1: Alphabetic disk in Ars Generalis Ultima*, 1308.*

---

365. Quoted by Henri Corbin, *Histoire de la Philosophie Islamique* (Paris, 1964), p. 205: "Sache que les secrets de Dieu et les objects de Sa science, les réalités subtiles et les réalités denses, les choses d'en haut et les choses d'en bas, sont de deux catégories: il y a les nombres et il y a les lettres. Les secrets des lettres sont dans les nombres, et les épiphanies des nombres sont dans les lettres. Les nombres sont les réalités d'en haut, appartenant aux entités spirituelles. Les lettres appartiennent au cercle des réalités matérielles et du devenir." Translation from the French, D.L.

Llull's motives are not easily understood today, but it seems safe to state that in his quest to convert the "infidels" to Christianity, the disk construction served theoretical functions as an encyclopaedia of religious thought, a tool to inspire meditation about its main topics, a way to generate new propositions, or even to abolish language altogether: "The idea is to present in a single book everything that can be thought [...] as well as everything that can be said [...]. From the binary combination of terms in this universal grammar, conceived as general principles, it would be possible to find a solution to any question the human mind can pose. As an art of questioning and getting answers to a variety of matters, it is applicable to all the sciences."[366]

Several authors have pointed out that the model for Llull's disks was a divinatory device called a *zā'irja*, which was in use among the Muslims he was seeking to convert.[367] The artefact offered a most astonishing function: Taking into account the moment in time of the enquiry, it generated a rhymed answer to any question posed. Llull was likely impressed by the fact that even kings placed a high level of trust in the procedure: "Many distinguished people have shown great interest in using [the *zā'irajah*] for supernatural information, with the help of the well-known enigmatic operation that goes with it."[368]

'Abd ar-Raḥmān Ibn Khaldūn's *Muqaddima* (1377), the introduction to a universal world history in seven volumes, presents a detailed and complete instance of the procedure executed on the "*Zā'irajah* of the world" (as one of the translators into English, Franz Rosenthal, spells it).[369] The device is attributed to the Maghribi Sufi Muḥamad b. Mas'ūd Abū l-'Abbas as-Sabtī, who lived in

---

366. Amador Vega, *Ramon Llull and the Secret of Life* [2002] (New York, 2003), p. 57. Fig. 1 p. 64.

367. Cf. Charles Lohr, Christianus arabicus, cuius nomen Raimundus Lullus. *Freiburger Zeitschrift für Philosophie und Theologie* 31. 1–2 (1984): 145–160; Dominique Urvoy, La place de Ramon Llull dans la pensée Arabe. *Catalan Review. International Journal of Catalan Culture* 4. 1–2 (1990): 201–220; D. Urvoy, *Penser l'Islam. Les Présupposés Islamiques de l'"Art" de Llull* (Paris, 1980); Armand Llinares, References et influences Arabes dans le Libre de contemplacio. *Estudios Lulianos* 24. 71 (1980): 109–127. The transliteration of the name varies considerably, "Zā'irjah," "Zā'irajah," "Zā'irdja," "Zā'iradja," "Zāyirǧa" being but a few examples. The Arabic letters in زايرجة are, from right to left: ز – zāy, ١ – '(alif), ي – hamza+i or yā`+i, ر – rā`, ج – jīm, ة – hā` (tā' marbūṭa). The name has tentatively been explained as a mixture of the Persian words for "horoscope, astronomical table" (zā'icha) and "circle" (dā'ira); cf. Ibn Khaldūn, *The Muqaddimah. An Introduction to History*, trans. Franz Rosenthal, 3 vols. (New York, 1958), vol. 1, p. 238f., fn. 364.

368. *Muqaddimah*, trans. Rosenthal, vol. 1, p. 239 [I, 213]. (The page numbers in square brackets refer to the Arabic edition of Quatremère.) The many surviving Arabic manuscripts by various authors attest to the popularity the device once enjoyed; cf. the catalogue of the Princeton Collection of Islamic Manuscripts (New Series and Garrett Collection, Yahuda Series), which list over 25 manuscripts on the topic; online: http://rbsc.princeton.edu/topics/1213/divisions/15.

369. "Muqaddima" signifies "Prolegomenon, introduction" in Arabic. The main printed editions of this work are, in chronological order: Ibn Khaldūn, *Kitāb al-'Ibar wa-dīwān al-mūbtadā' wa-l-ḫabar etc.*, ed. Naṣr al-Hūrīnī (Bulaq near Cairo, 1857, 1 vol.; reprint Bulaq, 1867/8, 7 vols.; in Arabic); *Prolégomènes d'Ebn-Khaldoun*, ed. Étienne M. Quatremère, 3 vols. (Paris, 1858; in Arabic); *Prolégomènes Historiques d'Ibn Khaldoun*, trans. William MacGuckin de Slane, 3 vols. (Paris, 1863–1868); Ibn Khaldūn, *al-Muqaddimah*, ed. 'Alī A. Wāfī, 4 vols. (Cairo, 1957–1962; in Arabic); Ibn Khaldūn, *The Muqaddimah. An Introduction to History*, trans. Franz Rosenthal, 3 vols. (New York, 1958); Ibn Khaldūn, *Discours sur l'Histoire Universelle (al-Muqadimma)*, trans. Vincent Monteil, 3 vols. (Paris–Beyrouth, 1967/8); Ibn Khaldūn, *Le Livre des Exemples, Vol. 1: Autobiographie, Muqaddima*, trans. Abdesselam Cheddadi (Paris, 2002).

Marrakech at the end of the twelfth century.[370] A footnote in the first printed edition of the work in Arabic from Bulaq near Cairo relates that Ibn Khaldūn derived the knowledge about the artefact "from people who work with the *zā'irajah* and whom [he had] met."[371] An untitled manuscript by Sheikh Jamāl ad-dīn Abd al-Malik b. Abd Allāh al-Marjānī in the library of Rabat, Morocco, establishes it was he who introduced Ibn Khaldūn to the procedure in Biskra, now Algeria, in 1370/1371 (the year 772 AH).[372] Al-Marjānī had heard from the *qāḍī* (judge) of Constantine, today also Algeria (who in turn had allegedly been informed by one of the companions of the prophet, H'udhaifa b. Yamān), that the *zā'irja* was a traditional and ancient science.[373] When Ibn Khaldūn questioned the correctness of the proposition, the sheikh suggested simply to ask the device itself:

الزايرجة علم محدث أو قديم

"*Az-zā'irja 'ilm muḥdāth au qadīm*" –
"[Is the] *zā'irja* [a] recent or [an] ancient science?"[374]

When the self-referential question was posed, the ascendant (the sign of the zodiac cycle rising at the eastern horizon) stood in the first degree of Sagittarius. Al-Marjānī performed the complicated procedure and explained it to Ibn Khaldūn. It yielded the answer "The Holy Spirit will depart, its secret having been brought forth / To Idrīs, and through it, he ascended the highest summit" ("*Tarūḥanna rūḥu l-qudsī ubriza sirruhā / Li-Idrīsa fa-starqā bi-hā murtaqā l-'ulā*") and thus confirmed the information the sheikh had given. The Qur'ānic sage Idrīs is commonly identified with the biblical Enoch and fulfils a function similar to the one described by Michael Thompson as "Father of all Tiv": the mythical ancestor at the absolute limit of memory.[375]

The manuscript from Rabat differs slightly regarding the time and reveals an unexpected trait of the learned lawyer (*faqīh*): "Then [after the answer had been obtained], [Ibn Khaldūn] started to dance and to turn on the terrace of his house. The ascendant of the question was the 18th [degree] of Sagittarius."[376] The difference of 17 degrees equals slightly more than one hour, since the sign at the Eastern horizon progresses by one degree every four minutes. If the time indicated refers to the moment when Ibn Khaldūn started to dance, the entire

---

370. *Muqaddimah*, trans. Rosenthal, vol. 1, p. 239 [I, 213] and fn. 365. Cf. Carl Brockelmann, *Geschichte der Arabischen Litteratur, Suppl. Vol. 1* (Leyden, 1937), p. 909.

371. *Muqaddimah*, trans. Rosenthal, vol. 3, p. 196, fn. 880.

372. Dates in this chapter generally follow the Christian system.

373. Henri P.J. Renaud, Divination et histoire nord-africaine au temps d'Ibn Khaldun. *Hespéris* 30 (1943): 213–221, esp. pp. 213–215. How the latter learned the fact from the former is unclear, since more than 600 years separate them.

374. The single letters of the question are: "'l z ' y r j t / 'l m / m ḥ d th / ' w / q d y m."

375. *Muqaddimah*, trans. Rosenthal, vol. 3, p. 213, fn. 921 [III, 178]; cf. Michael Thompson, *Rubbish Theory. The Creation and Destruction of Value* (Oxford, 1979), pp. 65–69.

376. Renaud, Divination, p. 215.

explanation took place in just 68 minutes, a very short time considering the operation's complexity.

Numeral Systems

The algorithmic artefact of the *zā'irja* – constructed from paper and driven by a human computer – processed symbols and was guided by rules. The symbols originated from three different alphabets, sets of signs that seamlessly converted between letters and numerals and belonged to two principally different ways of counting. During the period under consideration, from 1000 to 1400, two pre-positional and one positional system co-existed and the transition to the latter was about to take place.

The 28 symbols of the Arabic alphabet numerically represent units, tens and hundreds up to a thousand, without place value and zero (cf. Figure 2). They were called *ḥurūf al-jumal* ("letters for summing up, collecting"), or *abjad*, alluding to the four characters at the beginning.[377] The latter name derived from the mnemonics employed to memorise the numerical order of the signs, which was different from the alphabetic sequence after the symbols were rearranged by similarity in the seventh or eighth century. "Abjad" is the first of eight such words in Eastern Arabia, the others being "hawazin," "ḥuṭi-ya," "kalamuna," "s'afaṣ," "qurshat," "thakhudh" and "ḍaẓugh," which were later thought to represent archaic kings, demons, or the days of the week.

ا	ب	ج	د	ه	و	ز	ح	ط	ي	ك	ل	م	ن	ص	ع	ف	ض	ق	ر	س	ت	ث	خ	ذ	ظ	غ	ش
'	b	j	d	h	w	z	ḥ	ṭ	y	k	l	m	n	ṣ	'	f	ḍ	q	r	s	t	th	kh	dh	ẓ	gh	sh
1	2	3	4	5	6	7	8	9	10	20	30	40	50	60	70	80	90	100	200	300	400	500	600	700	800	900	1000

Fig. 2: Arabic letters, Western numerical interpretation.[378]

The *zimām* symbols follow the same system, but only designate numerals and do not contain a character for one thousand (Figure 3). They translate easily into Arabic letters of the same value. In both sets of signs, composite numbers are formed by joining together the respective symbols for units, tens, hundreds, and so on; in the first decreasing in value from right to left (the "natural" writing direction), and increasing in the second.[379] Hence 333 would be represented as سلج in *abjad*, or ﺳﻟﺞ in *zimām*.[380]

---

377. Edward W. Lane, *An Arabic–English Lexicon* [1863–1893] (Beirut, 1968), vol. 2, p. 459f. Cf. Georges Ifrah, *The Universal History of Numbers. From Prehistory to the Invention of the Computer* (London, 1998), p. 243f.

378. The Arabic Orient and Occident match letters and numbers slightly differently; cf. *Muqaddimah*, trans. Rosenthal, vol. 3, p. 173, fn. 809. Throughout this article, *alif* will be transliterated as ' and *'ayn* as '.

379. Azzedine Lazrek, personal communication, 14 May 2008.

380. The symbols employed in this paper, slightly different from the ones below, which Rosenthal copied from one of the manuscripts of the *Muqaddima*, were published by Azzedine Lazrek, Cadi Ayyad University, Marrakech, Morocco, as part of a proposal to include them in the Unicode standard; cf. A. Lazrek,

	1	2	3	4	5	6	7	8	9
	ل	ك	ٮ	٨	ٯ	ٮ	ٯ	ٯٮ	ٮ
	10	20	30	40	50	60	70	80	90
	ٮ	ٮ	ٮ	ٮ	ٮ	ٮ	٥	ٮ	ٮ
	100	200	300	400	500	600	700	800	900
	ٮ	ٮ	ٮ	ٮ	ٮ	ٮ	ٮ	ٮ	ٮ

333: ثلاث

Fig. 3: Zimām numerals, according to Rosenthal (1958).[381]

Georges S. Colin has shown that the genesis of the 27 signs can be traced back to one of the counting systems of antiquity. In addition to the 24 letters of their alphabet, the Greeks adopted three obsolete characters from their archaic script, ς (wau or digamma, = 6), φ (ḳoppa, = 90), and $\mathfrak{z}$ (ṣampi, = 900), and employed the resulting 27 symbols to designate units, tens, and hundreds. The Copts, Christianised Egyptians within the Byzantine Empire, used them from the third century onwards to calculate the finances of state. When Arabs conquered the country in 640, the Copts remained officials, but since the new governors of Egypt did not want the state finances exclusively in Christian hands, in 706 they obliged them to perform their work in the Arabic language and placed Muslim functionaries by their side. The Copto-Greek numerals that they continued to use probably entered the Islamic world via this route. It was not until the beginning of the nineteenth century that accounting with the esoteric numbers was finally forbidden in Egypt.[382]

In *zimām*, units, tens, and hundreds were designated by different characters and kept apart as qualitatively dissimilar. The highest value that could be represented by placing horizontal strokes below the signs was 999,999.[383] An example is appropriate here to illustrate the clumsiness of such systems: the addition of 222 and 444 in Roman numerals, CCXXII + CDXLIV, which is trivial in "Arabic" signs. First, a conventional hindrance, the subtractive notation in the second term, must be removed, which results in CCCCXXXXIIII. Next, the addition can be performed by grouping identical symbols together. The sorting produces CCCCCCXXXXXXIIIIII, which needs to be converted to standard form by reducing the signs to the highest values, DCLXVI, 666. Obviously, more complex arithmetical operations cannot be performed in this way, which is why they

---

Unicode – Proposals Symbols for Unicode Consortium, 17 March 2008; online: http://www.ucam.ac.ma/fssm/rydarab/unicode.htm. The proposal was accepted on 25 April 2008, and the characters will probably be allocated in the range 10E60..10E7E; cf. Unicode Consortium, Proposed New Characters – Pipeline Table, 1 May 2008; online: http://unicode.org/unicode/alloc/Pipeline.html.

381. *Muqaddimah*, trans. Rosenthal, vol. 3, p. 197, fn. 882.

382. Cf. Florian Cajori, *A History of Mathematical Notation, Vol. 1: Notations in Elementary Mathematics* [1928] (New York, 1993), pp. 21–30; Georges S. Colin, L'origine grecque des "chiffres de Fès" et de nos "chiffres Arabes." *Journal Asiatique* 222 (1933): 193–215, esp. pp. 193–203; cf. p. 202f., fn. 3: "It is odd to find that still at present, the majority of the functionaries of the Egyptian ministry of finances are Copts."

383. A. Lazrek, Rumi Numeral System Symbols (15 July 2006; online: http://www.ucam.ac.ma/fssm/rydarab/doc/unicode/n3087-1.pdf), p. 6.

were not executed with the numbers themselves, but with a separate artefact, the abacus. As only the results were written down, early Arabic mathematical treatises very rarely contain corrections and no intermediate calculations at all.[384] The device's speed is due to the mechanical performance of several simple operations.[385]

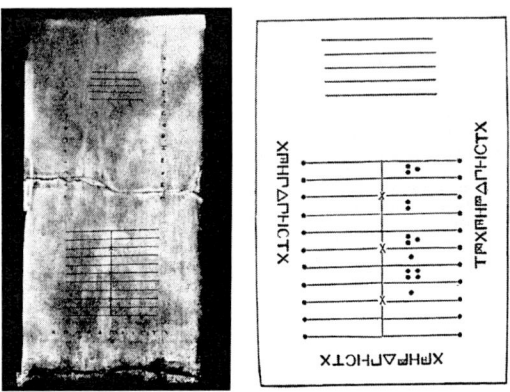

Fig. 4: Ancient Greek abacus found on Salamis, fifth to fourth century BC[386]

One of the main procedures on the reckoning-board is the scaling of values, where one bead is assumed to represent fives, tens, and so on. When a number needs to be carried over to the next counter, a second operation is employed, the calculation of the remainder after division through a certain quantity, 5 in the case of this device. It evolves naturally when a group of units needs to be compressed into a sign with a higher value, and may derive from the cyclical character of time and the practice of clock-building. The modern term is *modulo*. In Figure 5, the addition of 3 + 3 = 6 is performed on the abacus as an example. While the lower beads represent the units, the upper ones possess a value of five, and both are only

384. Cf. Mahdi Abdeljaouad, Le manuscrit mathématique de Jerba. Une pratique des symboles algébriques Maghrébins en pleine maturité, in: *Actes du 7ième Colloque Maghrébin sur l'Histoire des Mathématiques Arabes*, eds. Abdallah El Idrissi and Ezzaim Laabid (Marrakech, 2005; online: http://math.unipa.it/~grim/MahdiAbdjQuad11.pdf), pp. 9–98, p. 21.

385. Most basic calculations are executed faster on an abacus than with an electric calculator, as proven by a contest held in Japan shortly after the Second World War, in which the *Soroban* (Japanese abacus) champion won 4 to 1 against his North American counterpart; cf. Ifrah, *Universal History*, p. 289f.

386. The tablet measures approximately 149 × 75 × 4.5 cm. The right side of the device represents the number 9,823, 1 (×5000) + 4 (×1000) + 1 (×500) + 3 (×100) + 0 (×50) + 2 (×10) + 0 (×5) + 3 (×1); cf. Cajori, *Mathematical Notation*, p. 22f., and Ifrah, *Universal History*, pp. 201–203, for a more exact treatment. A similar table is already mentioned in the Athenian Constitution (written between 332 and 322 BC): "And when all have voted, the attendants take the vessel that is to count and empty it out on to a reckoning-board ["ábaka"] with as many holes in it as there are pebbles, in order that they may be set out visibly and be easy to count, and that the perforated and the whole ones may be clearly seen by the litigants. And those assigned by lot to count the voting-pebbles count them out on to the reckoning-board ["ábakos"], in two sets, one the whole ones and the other those perforated." *Aristotle in 23 Volumes, Vol. 23: Athenian Constitution*, trans. H. Rackham (Cambridge, MA, 1952), sect. 69. 1.

counted if they touch the centre. If three is to be added to the three entities already there, five needs to be carried over to the upper part, because there are not enough of them in the lower. After the operation, one is left in the units – the remainder when six is divided by five.[387]

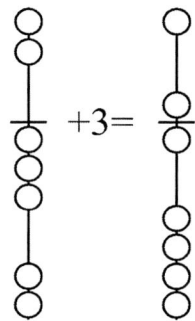

Fig. 5: Execution of addition on the abacus.

In contrast to *zimām*, the *ghubār* system represents numbers of any magnitude by arranging only ten signs from 0 to 9 (including zero) in a positional manner, which is comparable to the system most probably invented by Indian mathematicians and still employed today.[388] 333 would be expressed as ٣٣٣ – economically superior and mathematically momentous – using the same character to indicate units, tens, and hundreds.

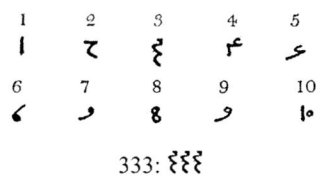

Fig. 6: Ghubār numerals, according to Rosenthal.[389]

The German naturalist and explorer Alexander von Humboldt had already pointed out in the nineteenth century how close the *ghubār* numerals were to the modern positional system by quoting his "friend and teacher" Sylvestre de Sacy, who had discovered them in a manuscript in the library of the old Abbey of St. Germain des Près: "The gobar has a strong relation with the Indian number, but it does not possess a zero."[390] Humboldt, however, believed zero was expressed by marks placed above the signs: tens were indicated by one, hundreds by two, and thousands by three dots in

387. Cf. Ifrah, *Universal History*, pp. 291–294.

388. The origins of the system in India is supported by many documents, but is by no means undisputed; cf. Ifrah, *Universal History*, pp. 356–439. For a controversial position, see Solomon Gandz, The origin of the Ghubār numerals, or The Arabian abacus and the articuli. *Isis* 16 (1931): 393–424.

389. *Muqaddimah*, trans. Rosenthal, vol. 3, p. 197, fn. 882. For the various forms of the numerals in the course of their history, see Ifrah, *Universal History*, pp. 534–539. According to Ifrah, "the oldest known documents which refer to *Ghubar* numerals and calculation date back to 874 and 888 CE" and were apparently found in India (p. 536).

390. Translation D.L. De Sacy also translated parts of the *Muqaddima*; cf. Baron Antoine Isaac Silvestre de Sacy, Extraits de Prolégomènes d'Ebn Khaldoun, in: *Relation de l'Égypte, par Abd-Allatif, Médecin Arabe de Bagdad*, trans. S. de Sacy (Paris, 1810; reprint Frankfurt a. M., 1992), pp. 509–524 (translation vol. 5, chap. 4 and vol. 4, chaps. 3 and 4), pp. 558–564 (Arabic text), and 'Abd-Ar Rahmān Al-Jāmī, *Vie des Soufis ou Les Haleines de la Familiarité*, trans. S. de Sacy (Paris, 1831; reprint Paris, 1977), pp. 16–20 (Arabic text), pp. 20–27 (translation "Du Sofisme").

superscript, constituting a "hybrid" counting system.[391] He speculated that it was these marks that had given the numbers their "strange name *gobar* or dust writing."[392]

Literally, "ghubār" (غبار) does indeed signify "dust," not in the sense of the miraculously appearing dust bunnies of our times, but rather the finest possible kind of sand.[393] This led many scientists in the nineteenth century to believe that the name derived from an ancient means of calculation. Since antiquity, a board of wood covered with dust (called "lawha" in Maghreb) had been used for this purpose, writing numbers with a stick on it and deleting others in a performative kind of arithmetic.[394] It complemented the wax tablets of Aristotelian fame. This setup allowed relatively flexible processing of data, unlike working with clay or stone, and could be described as volatile "random access memory," since its fluidity permitted the modification of any grain on the surface at any time.[395]

Conversely, Solomon Gandz showed in the 1930s that *ḥurūf al-ghubār* did not indicate "letters of dust," but rather "signs employed on the abacus," and that the word consciously reflected the etymology of Greek 'άβαξ derived from Hebrew אבק (abaq – "dust").[396] The first implementation of the technical form allegedly consisted of lines in the sand on a tablet and pebbles moved around on them, hence the name.

The *Muqaddima* of 1377 is the earliest document that mentions the utilisation of *ghubār* in North Africa. Apart from proving that the symbols were known by Muslim fortune-tellers, Ibn Khaldūn states they were "used for numerals by government officials and accountants in contemporary Maghrib," probably already for some time.[397] Only in the sixteenth century did the system become more common in Morocco, probably via Spain. Colin argues that even if the *ghubār* numerals are found in the Muslim Occident from the middle of the tenth century, their use was limited to mathematical calculations of Indian-type arithmetic. In everyday situations, the non-positional symbols were retained, and the two

---

391. In this representation, the number is decomposed into units, tens, hundreds, and so on, and then each is written down in two parts in a multiplicative manner. 7,659 would be expressed as 7^{000} 6^{00} 5^0 9, $7 \times 1000 + 6 \times 100 + 5 \times 10 + 9 \times 1$. The abbreviation of such systems led to positional notation very early on; cf. Ifrah, *Universal History*, pp. 330–340.

392. Alexander von Humboldt, Über die bei den verschiedenen Völkern üblichen Systeme von Zahlzeichen und über den Ursprung des Stellenwerthes in den indischen Zahlen. *Crelle's Journal für die reine und angewandte Mathematik* 4. 3 (1829): 205–231, esp. pp. 213, 222–224. In the manuscript that Rosenthal took the *ghubār* numerals from, zero is placed to the right side of the sign; cf. the ten in Figure 6.

393. Lane, *Arabic–English Lexicon*, vol. 6, p. 2224.

394. Abdeljaouad, Manuscrit mathématique, p. 20; Ifrah, *Universal History*, pp. 207–209. To learn how computation was performed on it, see pp. 556–560.

395. Real volatile storage only becomes possible when the medium employed is even more malleable, like water or electricity; cf. Chapter 3.

396. Gandz, Origin of Ghubār, p. 395f.; cf. Ifrah, *Universal History*, p. 207.

397. *Muqaddimah*, trans. Rosenthal, vol. 1, p. 239 [I, 214].

systems coexisted for several hundred years.[398] At the end of the seventeenth century, the *ghubār* began to replace Greek numbers, and seventy years ago, they were only still employed in one city of Morocco, Fes, among notaries. Although the distribution of inheritances was calculated in a positional system, it was fixed in *zimām*, in the hope that the "quasi-cryptographic" symbols would prevent the manipulation of the result by laymen.[399]

From the moment a positional system that includes zero is employed, the abacus is *aufgehoben* (annulled, conserved, and elevated) in the numbers. The technical form is transcribed into a more abstract and flexible one, which unlike previous schemata enabled the explosive development of arithmetic.

Material Forms of Thought

The operational basis of the *zā'irja* consists in three parts: two tables and a poem. The first and main element, the "front," shows a zodiac circle divided into eight concentric subsections with twelve chords (*awtār*) extending from the centre outwards, which carry signs from the different alphabets (see Figure 7 for a transliteration).[400]

A rectangular table of 55 vertical squares times approximately 131 horizontal squares, most of them also marked with symbols from the three alphabets, forms the second part of the system, the "back."[401]

---

398. Colin, Origine grecque, pp. 208–210. This appears less astonishing when one considers that the impractical Roman signs are still used today when calculations are not intended; for example, in the numbering of book volumes.

399. Colin, Origine grecque, pp. 194–196, 210. The Moroccan lawyer Abdelhamid Benmakhlouf kindly informed me that today, the numbers are still used by the Hebrew notary's office in his country (personal communication, 22 July 2008).

400. Facsimiles of the *zā'irja* front table, the back, and the concluding list from a Turkish manuscript can be found in *Muqaddimah*, trans. Rosenthal, and in *Histoire Universelle*, trans. Monteil.

401. The number specified in the text, 131, differs from the actual layout of the artefact. Rosenthal counted the quantity of horizontal squares in the Turkish manuscripts as 128 (*Muqaddimah*, trans. Rosenthal, vol. 1, p. 240, fn. 370), whereas the one he presented as a facsimile and also the one in the Cairo edition of 1957–1962 contains 129 columns (vol. 4, p. 1168).

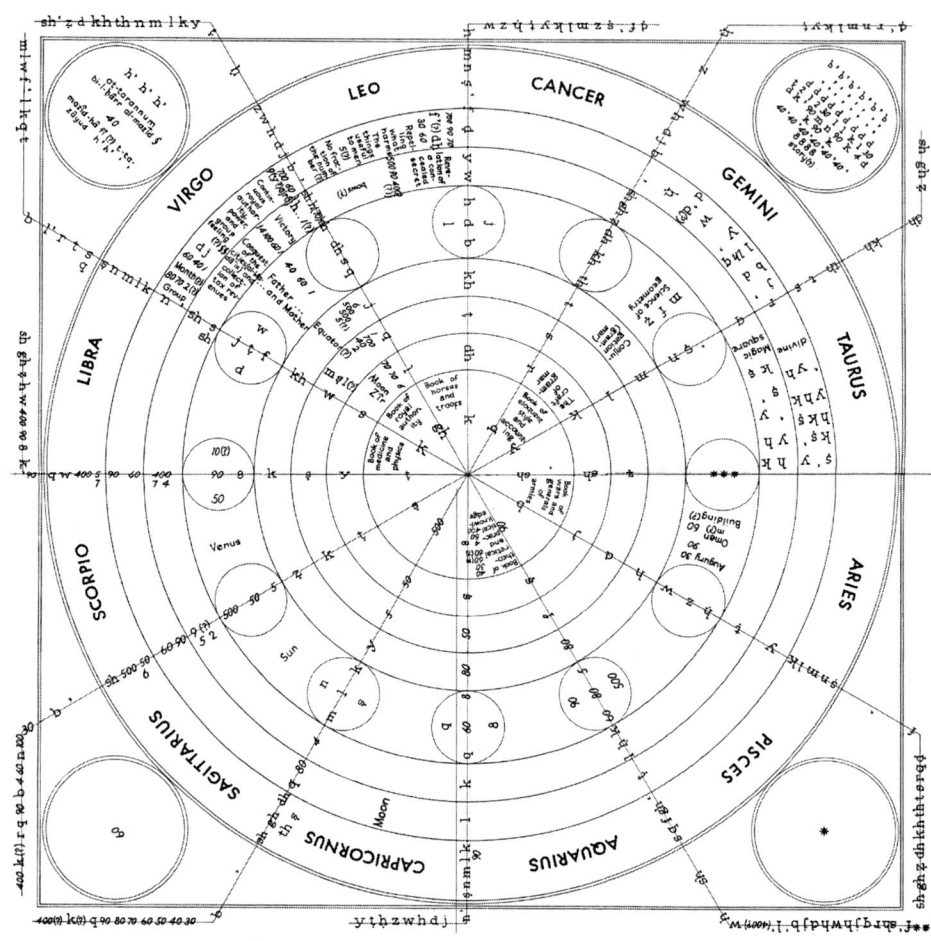

Fig. 7: Transliteration of the zā'irja front by Rosenthal.[402]

---

402. *Muqaddimah*, trans. Rosenthal, vol. 3, end pocket.

The third artefact needed to perform the procedure is an immaterial form of thought. The "key poem," ascribed to the Sevillian scholar Mālik b. Wuhayb, is entirely composed in the signs of one alphabet, the Arabic letters.[403]

سؤال عظيم الخلق حزت فصن اذن   غرائب شك ضبطه الجد مثلا

Rosenthal translated the signs "s w ' l / ' z y m / ' l kh l q / ḥ z t / f ṣ n / ' dh n // gh r ' y b / sh k / ḍ b ṭ h / ' l j d / m th l ' " – "su'al aẓym alkhalq ḥazat faṣun odhun / gharāyb shak ḍabṭuhu eljid mathalan" as "A weighty question you have got. Keep then, to yourself / Remarkable doubts which have been raised and which can be straightened out with diligence."[404] His successor Abdesselam Cheddadi rendered the poem as follows: "You possess the question of the grand natural form. Thus conserve the strange doubts that have been raised and which the diligence can dissipate."[405] The verse seems especially difficult to translate because it only represents a mnemonic aid, like the *abjad* words above, to remember otherwise meaningless sequences of characters. In Arabic chronograms, poems whose letter values reveal the date of an important event, a similar technique has reached the realms of literary art.[406]

An Archaeology of Algorithms

At the beginning of the detailed description of the procedure on the *zā'irja* in the third volume,[407] Rosenthal wrote in a footnote:

"The letters evolved in the procedure described by Ibn Khaldūn are marked in this translation by boldface type. However, the rationale of their determination and the relationship of the description to the table are by no means clear to me. As in the case of the *zā'irja* poem, a translation – one might rather call it a transposition of Arabic into English words – is offered here in the hope that it may serve as a basis, however shaky, for future improvement."[408]

---

403. Mālik b. Wuhayb lived from 453–525 AH, the early eleventh century; cf. *Muqaddimah*, trans. Rosenthal, vol. 1, p. 240, fn. 372; Seyyed Hossein Nasr and Oliver Leaman, eds., *History of Islamic Philosophy, Part I* (London, 1996), p. 296f.: "Among the logicians of Andalusia was Mālik ibn Wuhayb, famed for his learning in many sciences, including astronomy and (judicial) astrology, and known in his day as the Philosopher of the West." He later "turned his talents to divinity" and gave up on "open discussions of philosophy [...] because of the attempts on his life."

404. *Muqaddimah*, trans. Rosenthal, vol. 3, p. 211 [III, 176].

405. "Tu possèdes la question de la grande forme naturelle. Conserve donc les doutes étranges qui ont été soulevés et que le zèle peut dissiper." Cheddadi, *Livre des Exemples*, p. 1008. Translation from the French, D.L.

406. Cf. Mehr Afshan Farooqi, The secret of letters. Chronograms in Urdu literary culture. *Edebiyāt* 13. 2 (2003): 147–158; Ifrah, *Universal History*, pp. 250–252.

407. There is a short summary in the first volume, *Muqaddimah*, trans. Rosenthal, vol. 1, pp. 238–245 [I, 213–220]. The extensive account is found in vol. 3, pp. 182–214 [III, 146–179].

408. *Muqaddimah*, trans. Rosenthal, vol. 3, p. 197, fn. 880. The *zā'irja* poem Rosenthal refers to, which is attributed to as-Sabtī, precedes the detailed discussion of the device in the third volume. The complete obscureness of this long passage is probably due to the employment of some esoteric or cryptographic sort of code.

Rosenthal seems to have devoted considerable time and effort to understand the operations, as advocated by various footnotes tentatively trying to establish more consistency. Unfortunately, his papers have not yet been made accessible to the public.[409] Vincent Monteil, the translator of the French edition, commented:

> "The long 'explication' given by Ibn Khaldūn is – at least for us, humans of the twentieth century – remarkably confused. I have engaged myself in the succession of Rosenthal, profiting from his commendable efforts and the recourse to the tables that Slane could not have consulted. But the difficulties of translation are at times insurmountable, at least with the current state of our knowledge."[410]

Rosenthal employed the classical tools of philology to reconstruct the exact wording of the passage in question and the letters on the tables. Due to a less cautious attitude, Monteil again slightly corrupted the text in his translation.[411] Both failed to establish the concrete procedure executed on the device. The "Grand-Druide" Gwenc'hlan Le Scouëzec, a collector of practical knowledge of all kinds of esoteric systems, wrote in the "Dictionnaire des Arts Divinatoires": "The usage of Zā'irja consultation seems to have vanished, or, at least, to have been completely forgotten in the Maghreb. [...] Thus, it seems that the Zā'irja has fallen into more or less complete desuetude."[412]

Yet this pessimistic statement does not seem to be entirely justified. An Internet search of the letter sequence زايرجة (zā'irja) produced 804 results, among which, apart from Arabic editions of the Muqaddima, were several postings in newsgroups. These results show that at least some variant of the procedure is still used, for example, to calculate the answer to the question "Will America strike Iran this year?," which was computed in 2007. The answer was: "Yes, they will strike, but America will thereby be destroyed." The second enquiry, "When will America strike and who will win the war?" yielded "The war nears, the winner will be the Persians" (Iran today).[413]

The present chapter will attempt to rely on the "shaky ground" that Rosenthal had the merit of establishing, and to reconstruct the routine employed to generate the answer. The difficulty of the classical disciplines of historiography and philology to deal with the complexity of algorithms, and the devices within which they are embodied, demonstrates that their archaeology necessitates a different approach. The methods of cryptology offer an important aid in the reconstruction of regularity in mutilated or scrambled symbolic systems. Their applicability to a wider range of questions can be seen from the fact that they

409. After Rosenthal's death, they were given to Sourasky Library, Tel Aviv University, and have not yet been classified and catalogued (Miri Lipstein, director of the library, personal communication, 4 May 2008).

410. Muqaddima, trans. Monteil, vol. 3, p. 1123, fn. 1. Translation from the French, D.L. The text is in fact not confused; its complexity is confusing.

411. Cf. fn. 452 below.

412. Henri Veyrier, Encyclopédie de la Divination (Paris, 1982), p. 153. Translation from the French, D.L.

413. http://www.el7akeem.com/vb/showthread.php?p=28766, a website registered in Egypt. Other domains the word appears on are located in Saudi Arabia (4), Egypt (3), Morocco (1), Lebanon (1), and Syria (1). The author is indebted to Hicham Kerrouri and Samir Awaragi for translating these and other passages from Arabic.

are increasingly employed in genomics.[414] Through an analytical effort, an archaeology of algorithms and their artefacts aims at winning back as much ordered structure as possible from entropy, like the general discipline it belongs to.

The Signs of the Moment

In the first step of the procedure, the letters are taken off three of the chords on the *zā'irja*'s front, depending on the ruler and degree of the ascendant at the moment of the question. The sign rising on the eastern horizon wanders through the 360 degrees of the zodiac circle within a day. At the time, the data might have been obtained by consulting astronomical tables or by employing an astrolabe, an instrument closely resembling the one discussed here.[415] The description in the *Muqaddima* reads: "The ascendant is in the first degree of Sagittarius. Thus, we place the letters of the chord of the beginning of Sagittarius and the corresponding chord of the beginning of Gemini and, in the third place, the chord of the beginning of Aquarius up to the limit of the centre."[416] The degree and, consequently, the letters selected change every four minutes. Because the ascendant in the moment of the operation was Sagittarius, the characters are taken off the respective chord, following it over the centre to the opposite, Gemini. Then, the symbols bordering the third sign from the ruler are noted.[417] Unfortunately, Ibn Khaldūn does not mention if and how this relates to the current degree.

Even though the procedure at first sounds clear and simple, one encounters considerable difficulties to match the 73 letters listed in the text as "ṣ, ṭ, d, ṭ, h, n, th, k, h, m, ḍ, ṣ, w, n, th, h, s, alif, b, l, m, n, ṣ, 'ayn, f, ḍ, q, r, s, y, k, l, m, n, ṣ, 'ayn, f, q, r, s, t, th, kh, dh, ẓ, gh, sh, ṭ, k, n, 'ayn, ḥ, ṣ, z, w, ḥ, l, ṣ, k, l, m, n, ṣ, alif, b, j, d, h, w, z, ḥ, ṭ, y" with the chords they originate from.[418] This is due to several ambiguities and complications, which apparently none of the translators were able to overcome. First, it is unclear if the "beginning" of a sign is the chord on the left or the right side. Second, the text seems to imply that the letters are read following Sagittarius inwards over the centre, then along Gemini outwards,

---

414. Cf. Andrzej K. Konopka, Sequences and codes. Fundamentals of biomolecular cryptology, in: *Biocomputing. Informatics and Genome Projects*, ed. Douglas W. Smith (San Diego, 1994), pp. 119–174, p. 122f.: "Several methods for determining classification codes in large collections of nucleotide sequences are described in this survey. All these methods are analogous to techniques for [...] breaking monoalphabetic substitution cryptosystems [...]. In case of cryptology proper, the structure of 'functional messages' is known because the plain language is known. [...] The semantic aspects of 'functional messages' [in genomics], however, remain unknown even in principle because we do not know the 'language of Nature' nor is it obvious that such a language exists."

415. Cf. Arianna Borrelli, *Aspects of the Astrolabe. "Architectonica Ratio" in Tenth- and Eleventh-Century Europe* (Stuttgart, 2008), p. 65: "At the same time, though, the astrolabe was also a tool for astrologers, and astrology played an important role in tenth-, eleventh- and twelfth-century Arabic–Islamic court life."

416. *Muqaddimah*, trans. Rosenthal, vol. 3, p. 199 [III, 163].

417. It is actually the second, because the starting field is counted as 1.

418. *Muqaddimah*, trans. Rosenthal, vol. 3, p. 211 [III, 176].

and finally Aquarius inwards again. Third, it is not easy to decide which of the numbers and letters on the table are to be considered, some being located close to the chords, but not on them. Additionally, the mixing of the different sign systems adds to the uncertainty and creates an extra layer of obscurity. Several characters can be regarded as Arabic letters or as numerals of one of the alternative conventions, as their forms are sometimes very similar.

The difficulties do not primarily stem from mutilation in the course of historical transmission, which relied for a long time on hand writing. Rosenthal, who consulted most of the Turkish versions of the work, commented:

> "All these manuscripts have the same textual value that, in the period after the invention of printing, would be ascribed to a book printed under its author's supervision. There may be occasional mistakes, but a carefully written manuscript usually compares favourably with a printed text. Most manuscripts of this type may be confidently regarded as authentic copies of the text, and any factual mistakes or miswriting they contain may be considered the author's own. Under these circumstances, we should expect the variant readings to be comparatively few and insignificant. Collation shows this to be, indeed, the case."[419]

It will be seen below that the section on the *zā'irja* actually contains errors of transcription, but also of interpretation. This is due to the very special character of the chapter in question, and the almost complete arbitrariness – the absence of discernable rules the signs would be regulated by – of its basis, the two tables. The words describing the tables are determined by a dictionary, which only assigns meaning to some of them and, consequently, allows reconstruction in case of necessity. If no such order is employed, the only remaining system of rules is the alphabet, which correlates single letters and numerals in a certain sequence. The situation is further aggravated by the astonishingly high complexity of the operation, given the period in which it was invented. Lewis Carroll once illustrated its effect in the following way: "'Can you do addition?' the White Queen asked. 'What's one and one and one and one and one and one and one and one and one and one?' 'I don't know' said Alice, 'I lost count.'"[420] Obviously the rather sceptical Ibn Khaldūn, who was courageous enough to judge the language of Qur'ān as defective,[421] was impressed by the procedure as a serious method of enquiry guided by strict rules:

419. A list of the manuscripts Rosenthal consulted can be found in *Muqaddimah*, trans. Rosenthal, vol. 1, p. lxxxviii. Rosenthal wrote that an "exhaustive utilisation of all the manuscripts" could be expected from a forthcoming edition of the *Muqaddima* by Muhammad Tawit at-Tanji, which unfortunately never appeared in print. However, the translation by Cheddadi mentioned in fn. 369 above is based on a large number of copies consulted.

420. Lewis Carroll, *Through the Looking Glass* (London, 1871), p. 150f.

421. Cf. *Muqaddimah*, trans. Rosenthal, vol. 2, p. 382 [II, 341f.]: "Arabic writing at the beginning of Islam was, therefore, not of the best quality nor of the greatest accuracy and excellence. It was not (even) of medium quality, because the Arabs possessed the savage desert attitude and were not familiar with crafts. [...] The men around Muḥammad wrote the Qur'ān in their own script, which was not of a firmly established, good quality. Most of the letters were in contradiction to the orthography required by persons versed in the craft of writing."

"Many people lack the understanding necessary for belief in the genuineness of the operation and its effectiveness in discovering the object of enquiry. They deny its soundness and believe that it is hocus-pocus. The practitioner, they believe, inserts the letters of a verse he (himself) composes as he wishes, from the letters of question and chord. He follows the described technique, which has no system or norm, and then he produces his verse, pretending that it was the result of an operation that followed an established procedure. This reasoning is baseless and wrong. […] In order to refute this […], it is sufficient for us (to refer to the fact) that the technique has been observed in operation and that it has been definitely and intelligently established that the operation follows a coherent procedure and sound norms."[422]

Fixing the Chords

The number of the signs from the chords mentioned in the text is 73, which can be reached in only two ways, assuming the information about the zodiac they originate from is not completely wrong: Either its beginning is located on the right, and all symbols are taken into account, which results in 73 characters. Or the left is looked up, then numbers have to be disregarded, because otherwise there are too many signs – 93. If only letters are counted the result is 72. To find out which of the alternatives is the correct one, and, consequently, which of the chords the characters originate from, a classical tool of cryptanalysis has been employed. A frequency count of the letters given in the text ("T") is compared to that of the two possible sets of symbols on the zāʾirja, the left-hand ("L") and the right-hand ("R") alternative.[423] Figure 8 presents the results.

	ʾ	b	j	d	h	w	z	ḥ	ṭ	y	k	l	m	n	ṣ	ʿ	f	ḍ	q	r	s	t	th	kh	dh	ẓ	gh	sh
T	2	2	1	2	4	3	2	3	4	2	4	4	4	6	7	3	2	2	2	2	3	1	3	1	1	1	1	1
L	4	3	3	2	4	3	1	2	2	2	4	4	2	2	3	3	3	1	5	3	2	1	2	1	2	1	3	4
R	2	3	1	2	3	2	1	4	3	2	5	4	2	7	7	1	1	3	5	2	1	2	3	1	1	2	1	2

Fig. 8: Calculation of deviation between chord letters in text (T) and on front table, left-hand (L) and right-hand (R) alternative.

Apart from the matching extremes for "n" and "ṣ" (marked in grey), the sum of the squares of deviation equals 79 for the first case, for the second only 34, and the variance per letter is 2.82 versus 1.21. This clearly indicates that the "beginning" of a sign is located on the right, following the direction of Arabic script and the orientation of the zodiac circle, which starts with Aries and moves counterclockwise until the last one, Pisces, is reached. Without frequency analysis, the similarity can hardly be spotted.

---

422. *Muqaddimah*, trans. Rosenthal, vol. 1, p. 243f. [I, 217f.]. In 1684 the Archbishop of Canterbury, John Tillotson, gave the following explanation of the expression "hocus-pocus" which appears in the translation: "And in all probability those common *juggling* words of *hocus pocus* are nothing else but a corruption of *hoc est corpus*, by way of ridiculous imitation of the Priests of the Church of *Rome* in their *trick of Transubstantiation*." John Tillotson, *A Discourse Against Transubstantiation* (London, 1685), p. 34.

423. For the count of letters on the chords, Rosenthal's transliteration (Figure 7) was used.

The letters are taken off the chords and compared to the sequence found in the text. Apart from a region in the middle, which corresponds to Gemini read outwards, not many characters match, only about 26% (marked in grey).

T	ṣ	ṭ	d	ṭ	h	n	th	k	h	m	ḍ	ṣ	w	n	th	h	s	'	b	l	m	n	ṣ
C	t^{400}	k	r	q	$ḍ^{90}$	b	ḍ^d	$ṣ^{60}$	n	q^{100}	l^{30}	b	'	sh	th^{500}	n^{50}	w^{6}	$ṣ^{60}$	$ḍ^{90}$	$ṭ^{9}$	$ḥ^{5}$	b^{2}	th^{500}

'	f	ḍ	q	r	s	y	k	l	m	n	ṣ	'	f	q	r	s	t	th	kh	dh	ẓ	gh	sh	ṭ	k	n	'
n^{50}	$ḥ^{5}$	ẓ	k	ṭ	ṣ	y	k	l	m	n	ṣ	'		q	r	s	t	th	kh	dh	ẓ	gh	sh	y	ṭ	ḥ	z

ḥ	ṣ	z	w	ḥ	l	ṣ	k	l	m	n	ṣ	'	b	j	d	h	w	z	ṣ	ṭ	y
w	h	d	j	n	'	ṣ	n	m	l	k	$ḍ^{90}$	l	k	q	$ṣ^{60}$	$ḥ^{8}$	q	$ḥ^{8}$	f^{80}	n^{50}	ṣ	$ḥ^{8}$

Fig. 9: Comparison of symbols in text (T) and from chords (C), read inwards, outwards, inwards.[424]

However, some sequences from above repeat below in the opposite direction, like "k, l, m, n, ṣ, alif" towards the end (framed with thicker lines). If all the chords are read from the centre outwards, the resemblance of the letter groups becomes much closer, 75.3%.[425]

T	ṣ	ṭ	d	ṭ	h	n	th	k	h	m	ḍ	ṣ	w	n	th	h	s	'	b	l	m	n	ṣ
C	ṣ	ṭ	k	ẓ	$ḥ^{5}$	n^{50}	th^{500}	b^{2}	$ḥ^{5}$	$ṭ^{9}$	$ḍ^{90}$	$ṣ^{60}$	w^{6}	n^{50}	th^{500}	sh		'	b	l^{30}	q^{100}	n	$ṣ^{60}$

'	f	ḍ	q	r	s		y	k	l	m	n	ṣ	'	f	q	r	s	t	th	kh	dh	ẓ	gh	sh	ṭ	k
ḍ^d	b	$ḍ^{90}$	q	r	k	t^{400}	y	k	l	m	n	ṣ	'		q	r	s	t	th	kh	dh	ẓ	gh	sh	$ḥ^{8}$	ṣ

n	'	ḥ	ṣ	z	w	ḥ		l	ṣ	k	l	m	n	ṣ	'	b	j	d	h	w	z	ḥ	ṭ	y
n^{50}	f^{80}	$ḥ^{8}$	$ṣ^{60}$	q	q	$ḥ^{8}$	k	l	$ḍ^{90}$	k	l	m	n	ṣ	'		n	j	d	h	w	z	ḥ	ṭ	y

Fig. 10: Comparison of letters in text (T) and from chords (C), all read outwards.

As can be seen from the question mark that Rosenthal put at the "k" on the outer end of Sagittarius (see Figure 7), in places he could not decipher the signs unambiguously. Since he did not understand the procedure, he was unable to employ its regularity to correct the mutilation of single letters and had no means at his disposal to verify his transliteration.

The Arabic alphabet consists of approximately only 14 different basic forms, which are further differentiated into the 28 letters by diacritics. Historically, the marks were put above the signs when it became impossible to tell them apart:

---

424. *Zimām* and *ghubār* numbers were translated to letters and in this case noted in superscript.

425. The matching process can be compared to "anagramming" in cryptology; cf. David Kahn, *The Codebreakers. The Story of Secret Writing* (New York, 1967), p. 103.

150

"Certain groups of letter-shapes in the original Semitic alphabet were so simplified in the development of Arabic script that their forms became wholly identical [...]. The degree of confusion created was such that a system of dots had to be introduced in writing Arabic to differentiate letter-shapes that had merged," around the eighth century.[426]

In cursive Arabic script, most of what Western authors would consider the symbol's body is disregarded and the smallest conceivable difference, single dots, serves to tell characters like "b" (ب) and "t" (ت) , "j" (ج) and "kh" (خ), "f" (ف) and "q" (ق) apart. Even if their sound relates the letters, as in the case of "d" (د) and "dh" (ذ), their numerical value (which matters just as much in the *zā'irja* procedure) disagrees, with the character د representing 4, and ذ ,700. An irregularity in the paper or the sprinkling of ink may easily cause a dot to appear, just as bleaching of the material might make one vanish. Such mutilations can be spotted and corrected when one is dealing with meaningful text, because the miswritten word does not exist. Knowledge of the language then allows one to suggest a close alternative from the limited set of possible well-formed sequences. Semantically meaningless expressions can only be protected against distortion by repetition or, in cases where they were generated programmatically, by reconstructing the rules guiding them. In this regard the doubling of character sequences can be interpreted as a "null"-algorithm.[427]

Another factor that contributes to the unintended transformation of signs in Arabic originates from the way in which the scribe wrote them on the scroll. Because of its fragility, and probably also to avoid smearing the fresh ink when he moved his hand further to the left, the scribe turned the papyrus 90 degrees counter-clockwise and proceeded from top to bottom. Especially with uncommon characters it could happen that he performed the required mental rotation incorrectly. This explanation has been proposed to account for the change of form of the Arabic numerals 2, 3, and 7 around the year 1000.[428]

The reconstruction of the chords the letters were taken from provides cryptanalytically, as it were, a "depth of two,"[429] and allows a correction of Rosenthal's transcription.

426. Geoffrey Sampson, *Writing Systems. A Linguistic Introduction* (Stanford, 1990), p. 95f.

427. Given a set of input symbols, a programme produces a certain output. If both are identical, the algorithm performs nothing.

428. Papyrus was widely used in the Islamic world even after the Arabs had acquired the technique of paper production from the Chinese at the end of the eight century; Ifrah, *Universal History*, p. 532f., p. 516.

429. In Bletchley Park jargon during the Second World War, the term indicated the reception of two messages that had been encoded with exactly the same key settings and thus permitted a break-in.

The Transliteration Re-read

In the following three Figures (Figs. 11–13), the chords have been copied from the facsimile of the *zā'irja* front table and all places are marked where they disagree with the sequence specified in the text.

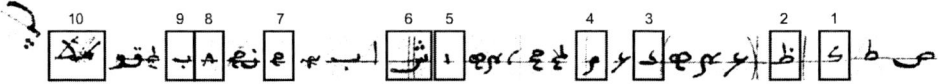

Fig. 11: Characters on Sagittarius.

The first differing letter, "k," ڪ (cf. Figure 11, read from right to left), appears as "d" in the text. Two of the manuscripts consulted by Rosenthal also read "k" at this place in the descriptive sequence. Two similar characters with the right value might have been miswritten, the Arabic letter "d," د, and the *ghubār* numeral 4, ۴. In general, signs in the text possess a higher weight of evidence because there are more copies of the manuscript than of the *zā'irja* table and because some of these symbols constitute the answer, which is a semantic (and thus rule-based) sequence. In this particular case, however, the letter does not appear in the output, and therefore, the most probable one could only be determined by a collation of all the versions of the passage. The second sign marked, "ẓ" (ظ), occurs instead of "ṭ" (ط). The difference between the two consists in a single dot that might have been added accidentally in the transmission process. Because of their context, Rosenthal interpreted the next two letters, ڪ and م, as numerals. In his transliteration, all the signs of the sequence underlined (symbols 5 to 15) were read as quantities: "5, 50, 500, 2, 5, 9, 90, 60, 6, 50, 500." Since he could not match the chords with the text, the only assumption left to him was that symbols from the different alphabets appeared in groups. The comparison of both sequences establishes that the signs in question should be read as letters, "k" and "m." He omitted symbol 5 altogether, a stroke half the size of the "alif" two positions later. The text suggests that it should be interpreted as a mutilated ه, "h." Regarding the next letter, again three dots above may have been added accidentally. Removing them results in "s" (س), the sign in the description. The seventh character was interpreted as *zimām* 100, ℮, while the explanatory passage shows the similar "m" (م) at this place, which might have been a typo.[430] It is part of the sequence "l, m, n, ṣ, 'ayn, f, ḍ, q, r, s" in the text, an excerpt of the Arabic alphabet in numerical order, from 30 to 300. In the transliteration of the *zā'irja*, it runs quite differently, "l^{30}, q^{100}, n, $ṣ^{60}$, d^4, b, $ḍ^{90}$, q, r, k, t^{400}." While in places an underlying logic allows the correction of transmission errors, it can also lead to the erroneous "reparation" of passages without order, and thus to mutilation of the document. It is impossible to say how the rather huge

---

430. The shape of "mīm" in Maghribi Arabic script published by Ifrah looks very similar to the sign encountered on the chord; cf. Ifrah, *Universal History*, p. 540.

difference at this location came about. Nor can an explanation be offered for the next character (8), which appears as *zimām* 4 in the transliteration of the chord and "'ayn" in the description. The next symbol could again have been transformed by a misplaced dot, and should be read as ف (f), as advocated by the sequence in the text. The last dissimilar sign (10) is hardly recognisable and Rosenthal even rendered it as two separate letters.

Since some of the characters from the chords are looked up in the course of the *zā'irja* procedure, they are repeated for a third time, which allows to further re-establish their original sequence. In particular, the last three letters of Sagittarius are chosen. It follows that there need to be exactly three signs at this location and that they must be "q," "r" and "s," otherwise the resulting answer would be different (for details of the lookup procedure see Figure 16, below). The mysterious last symbol was probably crossed out and corrected, since a "shin" (ش) was placed upside down at the location, a character similar to the required "s" (س, cf. Figure 11). In Arabic mathematics of the period, the letter, being an abbreviation of "shay" ("unknown"), usually indicated a free variable, much in the way "x" is employed today.[431]

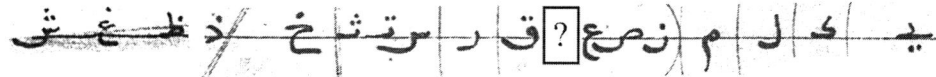

Fig. 12: Characters on Gemini.

Compared to Sagittarius, the symbols on the second chord are astonishingly well-preserved. This can be explained by the fact that they constitute an almost complete ordered sequence, the letters with the values from 10 to 1000 in the *jumal* system, excluding 80 and 90. *Zimām* or *ghubār* numbers that would need to be converted are absent. The only difference between the signs on the *zā'irja* and in the text is the appearance of an "f" at the place indicated with a question mark in Figure 12. This character is mandatory, as it is looked up during the procedure. The scribe might have mistakenly put down the two similar letters "f" (ف) and "q" (ق) by writing only the second on the chords and omitting the first under the misconception that he had already copied it.

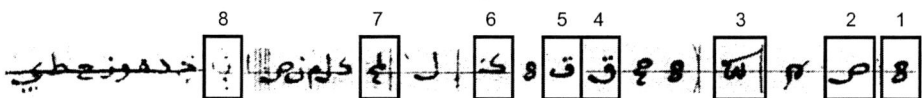

Fig. 13: Characters on Aquarius.

Turning now to the differences on the third chord, the first three cannot be accounted for. The next, "q" (4), might have resulted from an added second dot

431. Abdeljaouad, Manuscrit mathématique, p. 6.

on the "z" (ز) found in the description, in the same way that the following one, (5), might have been transformed by the accidental appearance of two dots on the "w" (و). The "k" (6, ک) is missing from the sequence in the text and needs to be removed for the procedure to work. A dot is found on its left, but remains meaningless because the letterform is never modified by diacritics in this position. The confusion of the 60 in the description and the 90 on the chords (7) may be due to the similarity of the uncommon *zimām* numerals, ٤ and ٤. The eighth mutilation shows a letter in the process of transformation. Rosenthal transcribed it as "n" (ن) because of the dot above it, but two are also located below. The comparison suggests that the only significant diacritic is the upper one of these, which results in "b" (ب). The *zā'irja* included in the Cairo edition of 1957–1962 supports the newly established characters at several places by providing the same ones, namely for Sagittarius 4, 7, and 9, and Aquarius 8. As can be seen in Figure 14, the above corrections increase the correspondence between the letters on the chords and those in the text to 93.2%.

T	ṣ	ṭ	d	ṭ	h	n	th	k	h	m	ḍ	ṣ	w	n	th	h	s	'	b	l	m	n	ṣ
C	ṣ	ṭ	k	ṭ	h^5	n^{50}	th^{500}	k	h^5	m	$ḍ^{90}$	$ṣ^{60}$	w^6	n^{50}	th^{500}	h	s	'	b	l^{30}	m	n	$ṣ^{60}$

'	f	ḍ	q	r	s	y	k	l	m	n	ṣ	'	f	q	r	s	t	th	kh	dh	ẓ	gh	sh	ṭ	k
$d^{ʲ}$	f	$ḍ^{90}$	q	r	s	y	k	l	m	n	ṣ	'	f	q	r	s	t	th	kh	dh	ẓ	gh	sh	$ḥ^8$	ṣ

n	'	ḥ	ṣ	z	w	ḥ	l	ṣ	k	l	m	n	ṣ	'	b	j	d	h	w	z	ḥ	ṭ	y
n^{50}	f^{80}	$ḥ^8$	$ṣ^{60}$	z	w	$ḥ^8$	l	$ṣ^{60}$	k	l	m	n	ṣ	'	b	j	d	h	w	z	ḥ	ṭ	y

Fig. 14: Comparison of letters in text (T) and from chords (C) after correction.

Except for the problems of interpretation mentioned above, the inconsistencies already exist in the Turkish manuscripts that Rosenthal consulted, one of which was signed by the author himself and evidently written under his supervision. We do not know how far the historian and learned lawyer Ibn Khaldūn was able to follow the intricacies of the procedure. It is said that letter magic takes a long time to learn and the Kabbalistic tradition goes so far as to demand that one should have reached the age of 40, or even 50, before studying it.[432]

---

432. Cf. Moshe Hallamish, *An Introduction to the Kabbalah* (New York, 1999), pp. 43–46.

The Numbers of the Cycles

After the characters have been taken off the chords, two numbers are calculated from the moment of the question, the first degree of Sagittarius, which in Arabic astrology is regarded as the fourth sign, counting from the end of the zodiac and not from its beginning.[433] The "principal cycle" (also called the "greatest cycle" or "greatest base"), which always equals 1, is added to the degree, 1, and the result multiplied by the "ruler": $(1 + 1) \times 4 = 8$.[434] If the number obtained is greater than 12, its modulo is taken. Since the degree ranges from 1 to 30 and the sign from 1 to 12, this first value lies between $(1 + 1) \times 1 = 2$ and $(30 + 1) \times 12 = 372$, but will always be scaled to fall within the interval of 1 to 12. The second number of the moment is calculated by adding the two astrological components, $1 + 4 = 5$, and will generally be located between 2 and 42. Now the reverse of the *zā'irja*, the rectangular table, comes into play:

> "The result of multiplying the ascendant and the greatest cycle by the ruler of Sagittarius [...] is entered at the side of eight. A mark is put upon the end of the number. The five that is the result of the addition of the ruler and ascendant is what is entered on the side of the uppermost large surface of the table. One counts, consecutively, groups of five cycles and keeps them until the number stops opposite the fields of the table that are filled" with "one of four letters, namely, *alif, b, j,* or *z*."[435]

The rightmost column of the table only contains eights, and this is why the text refers to it as the "side of eight."[436] The first value of the moment, 8, is entered here, moving from the bottom upwards. Then one counts in cycles of 5, the second number, until ا, ب, ج or ز is found, most likely from right to left. In the example, three of them pass before ا is hit. This is regarded as the result, once squared it amounts to 9, which constitutes the number of the first cycle.

The horizontal counting is very likely performed in the row reached by the vertical movement, and then, the top is looked up as "opposite," as the characters ا, ب, ج and ز make up most of it. While ب is written in normal orientation here, ج and ز are presented upside down as ﭗ and ﭕ, respectively, which adds to the obscurity. Using the table from the Turkish manuscript, the procedure cannot be consistently performed. However, the location of the sign can be deduced, because shortly afterwards 9 is entered in the same line, starting from the previous position:

---

433. Cf. *Muqaddimah*, trans. Rosenthal, vol. 1, p. 241 [I, 215]: "In the language (used here) the 'base' is the sign's distance from the last rank, in contrast to the (meaning of) 'base' in the language of the astronomers [?], where it is the distance from the first rank." The Arabic word in question is *uss*.

434. *Muqaddimah*, trans. Rosenthal, vol. 3, p. 199 [III, 163], and vol. 1, p. 242 [I, 216].

435. *Muqaddimah*, trans. Rosenthal, vol. 3, p. 200 [III, 164].

436. See note 400.

"The number in the first cycle, which is nine, must be entered in the front [...] of the table adjacent to the field in which the two [the 'side of eight' and the 'uppermost large surface'] are brought together, going towards the left, which is (the field of) eight. It thus falls upon the letter *lām-alif*, but no composite letter ever comes out of it. It thus is just the letter **t** – four hundred in *zimām* letters."[437]

Rosenthal remarked in a footnote that the numeral for 400, ﻄ, looks similar to the ligature *lām-alif*, ﻻ. In the eighth line, this sign is only found on positions 25, 26 and 27. Counting back 9 from this place, columns 16 to 18 are hit. The *ghubār* numeral 8 is located in the seventeenth, with an "alif" at the top, as required by the procedure. The section just quoted seems inconsistent, given it is the rightmost column of the table that is called the "side of eight." What is probably meant is that the counting starts next to the field that *contains* 8, going towards the left. How this location is reached "leaving three cycles behind" remains a mystery, especially because "alif" is already found five signs before, in column 12.

The complete operation takes place in twelve principal parts – as many as there are signs of the zodiac on the front of the *zā'irja*. Their numbers are calculated by adding the first value of the moment, 8, to the quantity obtained in the previous section of the procedure and taking its modulo 12 if it is greater than 12. In the example, this results in $(9 + 8 = 17; 17 \bmod 12 =)$ **5** for the second, $(5 + 8 = 13; 13 \bmod 12 =)$ **1** for the third, $(1 + 8 =)$ **9** for the fourth cycle, and then the sequence repeats. The periodicity depends on the two values of the stellar constellation. A start with 11 and an increment of 7, for example, leads to the unique series 11, 6, 1, 8, 3, 10, 5, 12, 7, 2, 9, 4.[438]

Within each of the twelve cycles, between two and six letters are calculated. Each half brings forth 20 symbols, the first triple of each, 9, and the second, 11, which exhibits a certain regularity. They are followed by three "results" that generate an additional seven. Accordingly, 47 letters are calculated in total (see Figure 15).

I(9)	II(5)	III(1)	IV(9)	V(5)	VI(1)	VII(9)	VIII(5)	IX(1)	X(9)	XI(5)	XII(1)	R1(9)	R2(5)	R3(1)
2	3	4	6	3	2	2	4	3	3	3	5	3	3	1
9			11			9			11			7		
15				11				14				7		
20						20						7		
47														

Fig. 15: Number of letters generated in each cycle, triple, quadruple, and first and second half.

---

437. *Muqaddimah*, trans. Rosenthal, vol. 3, p. 200 [III, 165], see also fn. 895.

438. This could be the outcome of the ascendant being in the 30th degree of the 5th sign, if seven cycles are counted before one of the signs is hit, $(30 + 1) \times 5 = 155; 155 \bmod 12 = 11$.

The following operations, which will be analysed in detail in the following sections, are performed in the course of the procedure:

– Lookup of letters in the signs of the three chords from the front of the *zā'irja*;
– Selection and conversion of symbols in the key poem;
– Generation of characters using the table on the reverse side;
– Scaling of values and miscellaneous arithmetic operations.

Chord Letters

In the selection of letters from the chords, the counting always starts from the beginning. The symbol hit is crossed out or otherwise marked and left out of subsequent calculations, which stochastically compares to an urn into which lots selected are not put back.[439] However, the procedure only works out as described if a certain number of letters is removed in every cycle. Since Rosenthal could not carry out the routine, he translated the deletion of signs in a way that is difficult to understand; only at the beginning are letters from the chords "dropped" (سقط – saqaṭ), which somehow indicates their removal, at all later places they are only "picked" (ضرب – ḍarab).[440] The characters are crossed out starting from the front, neglecting the ones that have already been looked up.

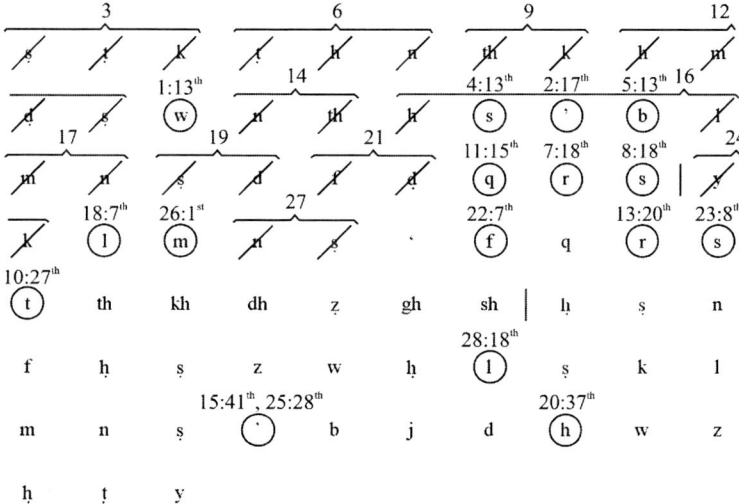

Fig. 16: Employment of chord letters.

439. Cf. the second problem in Jakob Bernoulli's *Ars Conjectandi*, J. Bernoulli, *The Art of Conjecturing, together with Letter to a Friend on Sets in Court Tennis* [1713], trans. Edith D. Sylla (Baltimore, MD, 2006), pp. 181–188.

440. They are "dropped" only at one instance in *Muqaddimah*, trans. Rosenthal, vol. 3, p. 202 [III, 166], and "picked" on pp. 203–210 [III, 168–175]; cf. Lane, *Arabic–English Lexicon*, vol. 4, p. 1379f. and vol. 5, p. 1777. According to this source, the first word signifies "drop" or "leave out," and the second "strike" or "hit."

Figure 16 depicts the lookup process in this part of the system for the complete procedure. Letters and strokes are numbered in the sequence of operations, and a second index shows the value leading to the sign. 13 is first converted with the help of the chords. The letters are counted from the beginning, and "w" in the second row is hit. In Figure 16 it was marked with 1, because this is the first step of the procedure, and with 13, the input value. A short time later, 17 is entered. Since "w" is now omitted, the "alif" in the second line is reached. It was annotated with 2 (second operation) and 17 (its position). Next, "[o]f the letters of the chords, one drops three."[441] The first three characters, ṣ, ṭ and k, are crossed out in the third step. Consequently, the 13 that is looked up next hits a different sign, "s" in the second line (marked with 4 and 13). The procedure in the chords can be comprehended by following the numbered order in Figure 16. If the letter sequence recovered earlier is used, the routine works completely as detailed in the text. The only inconsistency occurs at the fourth character in the seventh line, "alif." It is apparently looked up twice, which disagrees with the way the procedure is otherwise executed, which omits letters that were hit before. Seventeen symbols are generated from the chords, and 26 dropped. Figures 17 and 18 show their distribution in the different cycles, with no discernable regularity.

I	II	III	IV	V	VI	VII	VIII	IX	X	XI	XII	R1	R2	R3
w	'	s, b	r, s	t, q	r			l		h	f, s	', m		l
4				5				2			3		3	
6						4				4			3	
9								5					3	
17														

Fig. 17: Number of letters generated from the chords in each cycle, triple, quadruple, and first and second half.

I	II	III	IV	V	VI	VII	VIII	IX	X	XI	XII	R1	R2	R3
0	3	0	2, 1	2	2, 2	2	2	2	0	2	2	2	2	0
3				9				6			4		4	
6						10				6			4	
12								10					4	
26														

Fig. 18: Number of letters from the chords dropped in each cycle, triple, quadruple, and first and second half.

---

441. *Muqaddimah*, trans. Rosenthal, vol. 3, p. 202 [III, 166].

The Key Poem

In a footnote, Rosenthal wondered about the meaning of the frequently encountered formulation that letters produced were marked "(as belonging) to the verse of the poem."[442] In places a number is specified that is employed for this purpose. In the second cycle, one generates a "w" and "marks it with four [?] (as belonging) to the verse of the poem."[443] Some values are added to it, resulting in 17, which is converted back by looking it up in the chords, yielding "alif," as mentioned in the previous section (see Figure 16, second row, third to last sign). The key poem is employed as a fixed lookup table assigning each letter a number and vice versa, comparable to an *abjad* with a mapping different from the conventional one. The substitution alphabet can be used in both directions and represents the generative "motor" of the procedure. Characters produced are transformed back into different quantities, which in turn can be further processed. Converting 1 to "alif" and then "alif" back to 1 would, in contrast, only yield a monotonous sequence, ||||| .... The question remains as to why the "w" is annotated with 4, when it is only the second letter in the key poem. Figure 19 presents all characters with their position in the verse and their respective "marks." Where the text does not mention a number, the cell in the table has been left blank.

t^1	w^1	w^2	$\prime^1$	q^1	s^1	b^1	z^1	r^1	$\prime$	$\prime$	r^2	s^2	$\prime^2$	t^2	q^2	b^2	$\prime^4$	r^4	q^4	$\prime^8$	$ʿ$
16	2	2	3	13	1	27	15	24	25	34	24	1	3	16	13	27	3	24	13	3	5
		4		27	15						48	2	6	32	26	54	12	96	52	24	

$\prime^{16}$	r^1	$ḥ^1$	r^2	$ḥ^2$	l^1	d^1	r^4	s^4	h	l^2	d^2	y^1	s^8	r^4	$\prime^{32}$	m	t^4	$\prime$	l
3	24	14	24	14	4	37	24	1	33	10	37	7	1	24	3	8	16	9	4
48	24		48	28		37	96	4	5	20	74	7	8	96	96		64	(9)	

Fig. 19: Letters generated with real position in verse and "marks" from key poem.

The index that has been added to the characters solves the riddle of their "marking as belonging to the verse." It is a multiplicand called the "letter cycle," which starts from 1 and is doubled every time the respective sign is looked up again.[444] Symbols are "marked" with the product of their position in the key poem and this value. Several times in the procedure (framed in Figure 19) "r" is converted. At the first position, no number is mentioned, but the resulting quantity is doubled at the second position, quadruplicated at the third, and so on. The "w" was translated to 4 because it had been looked up before, at the second position. One can only speculate that every character was annotated with a value, and that the text only mentions it explicitly when the number is employed

---

442. *Muqaddimah*, trans. Rosenthal, vol. 3, p. 200, fn. 896.

443. *Muqaddimah*, trans. Rosenthal, vol. 3, p. 202 [III,166]. The question mark is Rosenthal's, D.L.

444. *Muqaddimah*, trans. Rosenthal, vol. 3, p. 209 [III, 174].

subsequently. The scale is reset when the "mark" equals or is greater than 96, as shown by a comment in the text when the first "r" in the lower part of the table is generated: "Having reached ninety-six, the whole thing starts from the beginning, which is twenty-four."[445]

This theory accounts for all quantities obtained with the help of the key poem. It only fails to explain why the "h" marked in grey in the lower part translates to 5, even though its position is 33 (the text seems to be corrupted here).[446] Additionally, the last "r" should have been converted to 24, not 96. Ibn Khaldūn commented: "Its sign is ninety-six, which is the end of the second cycle of letter cycles," but no remark is provided at the previous instance of the character, where it was already translated to 96 and should have been reset.[447] Of the "alifs" highlighted, the second and third follow a special procedure and are looked up at a later position in the poem: this is why the value is only doubled on the fourth occasion.

At the end of the section on the *zā'irja*, the letters of the verse are listed under each other with numbers written on their left.[448] Next to them, the twelve cycles with their respective start values are noted, which creates the misleading impression that they would somehow be guided by the characters of the poem. In fact, it was only written down in this way to facilitate its main function, the conversion of letters to numbers, and the proximity occurred accidentally. The term "key poem" should also be understood in the cryptologic sense of the term. It is not surprising that the end of the list was treated rather carelessly, as Rosenthal noted, since only the first appearance of a letter was required for the conversion to numerals. Of the four characters "written to one side," three already occurred earlier in the verse, and none of them are accessed in the course of the procedure.[449]

In Figure 20, positions that are used to generate numerals as "belonging to the poem" have been marked with indexes showing the order of events, as earlier. Less frequently, the verse is employed in the opposite direction, to convert quantities to letters. The respective locations have been circled and the superscript underlined. The characters are generally counted from the beginning and left in place afterwards. On two occasions the calculation continues from the last position looked up, in which case a "c" was added to the index.

445. *Muqaddimah*, trans. Rosenthal, vol. 3, p. 206 [III, 171].

446. Cf. *Muqaddimah*, trans. Rosenthal, vol. 3, p. 208, fn. 904. The *abjad* value of "h," however, is 5.

447. *Muqaddimah*, trans. Rosenthal, vol. 3, p. 209 [III, 174].

448. *Muqaddimah*, trans. Rosenthal, vol. 3, p. 212 [III, 177]. See note 400.

449. *Muqaddimah*, trans. Rosenthal, vol. 3, p. 213, fn. 916.

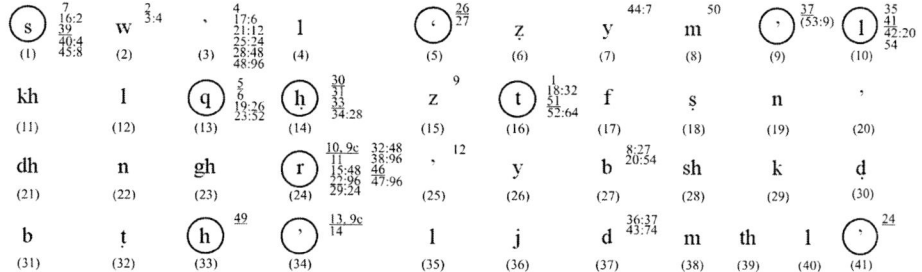

Fig. 20: Key poem with bi-directional operations marked.

The quantity of characters produced through the verse rises regularly by one in each triple, as can be seen in Figure 21.

I	II	III	IV	V	VI	VII	VIII	IX	X	XI	XII	R1	R2	R3
	q	r, '				(')[450]	', ḥ	ḥ	'	s, l	r	h	t	
	1			2			3			4			2	
	3				2				5				2	
		3					7						2	
						12								

Fig. 21: Number of letters generated from key poem in each cycle, triple, quadruple, and first and second half.

The Reverse Table

The lookup of values on the reverse table, the largest and most complicated artefact employed ($129 \times 55 = 7,095$ fields with several thousand characters), is very difficult to reconstruct.[451] At the beginning of each cycle, its basic number is entered at the side of eight, starting from the position in the last and counting upwards, and then all operations are carried out in the row reached. A detail permits reconstruction of the vertical movement with a relatively high degree of accuracy. In the tenth cycle, the text reads: "One goes up nine on the side of eight. There is an empty (field). One goes up another nine and gets into the seventh (field) from the beginning."[452] Adding up the preceding vertical steps, one gets: $8 + 5 + 1 + 9 + 5 + 1 + 9 + 5 + 1 = $ row 44. Going up nine from there, the pointer is located in line 53, counted from the bottom. Even if only three signs are found here, it is not completely clear why this row is

---

450. This value is not used as a letter, but represents the "scale of the second growth," one (*Muqaddimah*, trans. Rosenthal, vol. 3, p. 205 [III, 170]).

451 See note 400.

452. *Muqaddimah*, trans. Rosenthal, vol. 3, p. 206f. [III, 172]. By contrast, in the French translation the movement on the side of eight is often translated in an absolute manner: "On remonte au 5 sur le côté des 8" (*Muqaddima*, trans. Monteil, vol. 3, p. 1133).

considered empty. If one climbs up another nine, the pointer goes over the top of the table and, restarting to count at the bottom, stops in the seventh line from the beginning, in agreement with the text, $53 + 9 = 62$, $62 \bmod 55$ (the total number of rows) $= 7$.

I	II	III	IV	V	VI	VII	VIII	IX	X	XI	XII	R1	R2	R3
t	n, w	z	', '	b	'	q	', r	r	d, r		d, y		', l	
4			4			4			4			2		
6				5				5				2		
8						8						2		
18														

Fig. 22: Number of letters generated from reverse table in each cycle, triple, quadruple, and first and second half.

Figure 22 presents a count of the characters taken from the back of the artefact. Four of them are produced in each triple in a perfectly regular way. However, considerable difficulties were encountered in establishing the horizontal movement and precisely locating the signs mentioned. Values are entered in the "front," and then either the symbol upon which the number "falls" is taken, or the "surface" is consulted and the letter obtained from there. The following tentative remarks are written in the hope of providing some more "shaky ground" on which to base future decryption attempts. The most successful strategy was to only count cells filled with characters in the current row, omitting the field of the "side of eight" and each time restarting from the right side, unless explicitly indicated otherwise in the text.

In Figure 23 the procedure in the reverse table has been noted. The first column contains the number of the cycle, the second the vertical movement in the "side of eight," and the third, the horizontal movement on the "front." The fourth column contains the resulting location of the pointer, specifying first the x, and then the y position. The fifth column gives the correction that had to be applied in some cases to reach the character mentioned in the text. If the sign was taken from the current row of the table, it was entered in the sixth column, and when the "surface" was looked up, it was entered in the seventh. If both signs were mentioned, the one not employed in subsequent operations has been put in parentheses. Next follows the numeral value of the sign and any calculations applied before converting it to the letter that results, which is found in the last column.

cycle	ymov	xmov	location	corr.	table	surface	numeral	letter
I.	8		1/8		8			
		15 (3 × 5)	17/8		(8)	١	3	
		9c	26/8		ܥ		400	t
II.	5c		1/13		8			
		17 + 5	42/13		ܦ		500 = 50	n
		5	24/13		ܘ	١	5 + 1 = 6	w
III.	1c		1/14		8			
		13	34/14	(+1)		ܩ	3 × 2 + 1 = 7	z
IV.	9c		1/23		8			
		9	24/23			١	1	'
		11	27/23[453]			١	1	'
V.	5c		1/28		8			
		26	50/28		ܚ		2	b
VI.	1c		1/29		8			
		18	36/29	(+1)		١	1	'
VII.	9c		1/38		8			
		10	35/38	(−2)	ܦ		500 = 50, 50 × 2 = 100	q
VIII.	5c		1/43					
		5	20/43			١	1	'
		52	101/43	(−3)	ܚ		2 = 200	r
IX.	1c		1/44		8			
		52	101/44	(−2)	ܚ		2 = 200	r
X.	9c		1/53		%			
	9c		1/7		8			
		36	92/7		ܪ		4	d
(		9	17/7		8		8 / 2 = 4	)
(		18	27/7		ܘ		1 = 10; (10 − 2) / 2 = 4	)
(		27	40/7?		ܥ		(10 − 2) / 2 = 4	)[454]
		26	?		ܚ,ܖ,ܩ		200	r
XI.	5c		1/12		8			
		5 (4?)	5/12[455]		%	١	1	
XII.	1c		1/13		8			
		1	1/13		8		8 / 2 = 4	d
			?			5	5 + 5 = 10	y
R1	9c		1/22					
		9	?		(ܚ)	ܩ	3 × 9 + 7 − 1 = 33	
		18	?			١	1	
R2	5c		1/27					
		9 (39?)	?		(ܦ)	١	1	'
		9	?		ܠ		30	l

Fig. 23: Lookup process in reverse table.

---

453. The symbol reached by the previous 9 has to be omitted in counting.

454. These calculations are only performed in a hypothetical way.

455. At this place, empty fields are obviously counted.

To illustrate the notation used, the second cycle will be discussed as an example. At its beginning, the pointer at the side of eight, which is located in the eighth row, moves up 5, and consequently reaches field [1/13]. On the front, one enters 17 and then 5, finding the *zimām* symbol ٯ at [42/13]. Its value is 500, but it is scaled into the tens, 50, which corresponds to "n" according to the *abjad* order. Then, 5 is looked up in the same row, leading to field [24/13], which contains ٥, 5. Turning now to the surface, that is, the top of the column, one finds ١, 1. The values obtained are added, 5 + 1 = 6, and the result is converted to the letter "w."

Most of the signs required for the procedure can be found in the exact field that is reached. Operations that work as described in the text have been marked in grey in Figure 23. If a certain fuzziness is accepted, some more symbols can be located in the vicinity to the right or left of the correct spot. Distortion of values is all the more to be expected, since two complete columns are apparently missing from the table, which was said by Ibn-Khaldūn to measure 131 squares horizontally.

It can be seen that the relatively high success rate at the beginning decreases after the tenth cycle. Possibly, from this point on the vertical movement was executed differently than assumed. Two "fixpoints" may help to establish the location of the pointer towards the end. In the first "result," R1, the number 9 falls on *ghubār* 2, ٢, in the table, while the surface contains 4, ٤. Counting 9 in the corresponding row leaves the pointer in [38/22]. In the vicinity of this address, the coincidence of the two signs only happens in fields [46/27], [50/28] and [34/29]. In the second result (R2), the combination of *ghubār* 30, ٣, in the table and "alif" at the top, which should be located at [26/27], can be found at [24/28] and [23/28] to [23/31]. The author has been unable to determine a coherent procedure that would connect these coincidences.

The Complete Routine

The question "how to best describe algorithms" has yielded impressively diverse answers, going so far as to propose we rely on Critical Theory to perform the task.[456] The position of this chapter is simple: The best representation and basis for investigation is the most concise and least ambiguous one, which will most closely approximate its complexity.

---

456. Cf. Mark Marino, Critical Code Studies. *Electronic Book Review*, April 2006, online: http://www.electronicbookreview.com/thread/electropoetics/codology.

init		
zodiac: <u>4</u>, degree: 1, v1: <u>8</u>, v2: 5		

I (<u>9</u>)	II (<u>17</u>, <u>5</u>)	III (<u>13</u>, 1)
9Tm, 9+4, 13Cm	17+5, 22Ts, 5T+Sm<u>4</u>, 4+8+5, 17Cm, Cd3	13Pm, 13Cm, 1cCm<u>27</u>, 13S×2+1m15
t, w	n, w, '	q, s, b, z

IV (9)	V (<u>17</u>, <u>5</u>)	VI (<u>13</u>, 1)
Cd2, 9cPm, 9Sm, 9cPm, Cd1, 9×2, 18Cm48, 18Cm<u>2</u>, 9+2, 11Sm6	Cd2, 5×2+17, 27Cm3<u>2</u>, 17−2, 15Cm26, 26Tm5<u>4</u>	Cd2, 1+4+13, 18Sm1<u>2</u>, Cd2, 18+2, 20Cm96
r, ', ', r, s, '	t, q, b	', r

VII (<u>9</u>)	VIII (<u>17</u>, <u>5</u>)	IX (<u>13</u>, 1)
Cd2, 9+1, 10Ts×2m<u>52</u>, 52−2−9, 41Cm24, 41P	Cd2, 5Pm, 5Sm48, 48−1+5, 52Tsm24, 24+5−1/2, 1<u>4</u>Pm	13×4, 52Tsm48, 13+1, 14Pm28, 14−7, Cd2, 7Cm
q, ', (')	', ', r, ḥ	r, ḥ, l

X (<u>9</u>)	XI (<u>17</u>, 5)	XII (13, <u>1</u>)
9×4, 36Tm37, 9P, 9×3−1, 26Tm96	Cd2, 4S1Pm4, 17×2−1+4, 37Cm5, 5×2, 10Pm20	Cd2, 1T/2m74, 1S+5m7, 7C, 7+1, 8Cm8, 8×3, 24Pm96
d, ', r	s, h, l	d, y, f, s, r

init		
9<u>3</u> letters, 93 : 12 = <u>7</u>, 93 % 12 = <u>9</u>		

R1 (<u>9</u>)	R2 (17, <u>5</u>)	R3 (<u>13</u>, 1)
Cd2, 9×3+1, 28Cm96, 9S<u>3</u>, 3×9+7−1, 33P, 9×2, 18S1,1Cm	Cd2, 5×3+1, 16Pm64, 5+3+1, 9T30S(m9), 9Tm	13+3+1+1, 18C
', h, m	t, ', l	l

legend
T: lookup in table
S: lookup table surface
C: lookup chord letters
P: lookup in key poem
m: marked as belonging to the poem
s: scaled
d: dropped from chords

Fig. 24: Overview of the complete zā'irja procedure.

Figure 24 condenses the operation in its entirety into a sort of scripting language. In the first row, the cycle together with its number is found, in the second, the steps taken are encoded according to the legend, and in the third are the resulting letters. Arithmetic operations are noted without precedence, merely within the sequence they are performed. The procedure in the different cycles follows no discernable regularity concerning the order in which the different parts of the system are employed. In the first, the table is looked up, then the chords (T–C); in the second, table, surface, and then chords (T–S–C); in the third, poem, chords, chords, surface (P–C–C–S). No description of how the sequence of events is determined exists in the text. Therefore, it cannot be established whether it is simply irregular, or because of some rule it is conditionally dependent on certain previous results.

In the last step of the procedure, the symbols are shuffled by dividing the sequence into two parts, one of 24 and the second of 23 characters, between the "alif" and "r" in the eighth cycle, and then writing down the first letters of both, after that, the second ones, and so on. This basic columnar transposition rearranges the output to: "t, r, w, ḥ, n | r, w, ḥ | ', l, q, d, s | ', b, r, z | s, r, h, ' | l, ', d, r, y, s | f, ', s, t, r, q, ' | b, h, ', | m, r, t, q, ' | ', l, ', l, ','' which allows interpretation as an Arabic verse, "Tarūḥanna rūḥu l-qudsī ubriza sirruhā / Li-Idrīsa fa-starqā bi-hā murtaqā l-'ulā." – "The Holy Spirit will depart, its secret having been brought forth / To Idrīs, and through it, he ascended the highest summit." Since the shuffling makes it difficult to foresee the final location of the letters, it is hard to imagine that the practitioner composed the verses at will and merely simulated the formal procedure by picking calculations that would result in the necessary numbers.

The Transfer of Forms

In the course of the operations, values obtained are often added, as in the second part, where the "w" is marked with four "as belonging to the poem," and this is then summed up with eight, the first value of the moment, and five, the remainder of the cycle's number, resulting in 17, which is looked up in the chords and generates "alif."[457] Less frequently, subtraction is employed, as is the case in the seventh portion of the routine, where "q" is converted to 52 in the verse, and the two is dropped, without explanation, along with the nine that is the remainder of this section's principal value, resulting in 41, which is entered in the chords and yields "alif." The "q" itself was the outcome of another operation, that of scaling between units, tens and hundreds. One "enters ten on the front (recto) of the table, and gets thus to a stop at five hundred. It is, however, (counted) only as fifty, n." The text does not indicate how the scale is determined. The final letter is reached by yet another technique, the frequent doubling or halving of values: "It is to be doubled. Thus, it is **q**," 100 in *abjad*.[458] The most unusual operation occurs in the calculation with the *uss* (literally, "base"). While the term usually means the exponent of a term, in this context it seems to indicate the unit fraction of a composite number. In the fifth part, "t" is marked with 32, and then the practitioner "subtracts the two which is at the base of thirty-two, from seventeen," the cycle's principal value, and enters fifteen into the chords, generating "q."[459] Where the *uss* is taken, the unit number has been underlined in Figure 24, like all other output that is subsequently employed. In certain places, a value of one is added or subtracted in the course of the routine, in cycles 7, 8, 11 and 12, at the end of the second and third quadruple, and in all three results.

---

457. *Muqaddimah*, trans. Rosenthal, vol. 3, p. 202 [III, 166].

458. *Muqaddimah*, trans. Rosenthal, vol. 3, p. 205 [III, 170].

459. *Muqaddimah*, trans. Rosenthal, vol. 3, p. 203f. [III, 168]. For the explanation of *uss*, see fn. 898. The same term already appeared in a different function before, see fn. 433. The previous retention of 50 from 52 in the seventh cycle possibly represents a similar procedure with the tens. As can be seen in the sixth section, where the *uss* of 54 is 4, the meaning of the expression cannot be "smallest divisor."

The arithmetical procedures on the *zā'irja* relate to the abacus. As shown above, the modulo operation, which is called *taksīr* ("breaking down") in Arabic letter science, occurs when quantities have to be carried over on the calculation device.[460] In the instrument of divination, it is found on the front table, where the selection process cycles depending on the ascendant. The numbers of the principal parts are calculated employing modulo arithmetic and the consultation of the rectangular table, as well as of the key poem, restarts from the beginning when the end is reached. The scaling of quantities, one of the main properties of the abacus and closely related to the system of place value, also occurs on the *zā'irja*, as mentioned above. Equally, the retention of the *uss* is only possible if numbers are dissected into units, tens, hundreds, and so on. Humboldt wrote:

> "In the Orient, the negromantic [*sic*] art of sand is called *raml*. Continuous or broken lines and points, which represent the elements, guide the soothsayer. [...] [T]he parallel lines of the magic books, similar to notes, knotty, often broken, seem to be mere graphical projections of these **chords of calculation and thought** [which make up the abacus]."[461]

A number of similarities support the opinion that Llull constructed his paper machine inspired by the algorithmic artefact of the Muslims. Charles Lohr wrote that the *zā'irja* provided "many close parallels not only to the association of letters with the elemental powers in Llull's early Artes and to the application of them to astrology in the *Tractatus novus de astronomia*, but also to the triangles and revolving circles which have been responsible for much of the misunderstanding of Llull's intention in the Art."[462]

– The front table of the *zā'irja* and the Majorcan device are visually similar. In the divinatory artefact, a major part of the answer is determined by a cyclical revolution, the natural movement of the stars as seen from Earth, which also determines time. In Llull's *Art*, turning the disks in regard to each other effects the recombination of terms, an element that seems absent from the Arabic device. But here as well, some of the letters taken off the chords are selected and then shuffled with characters that originate from other sources.

– The *zā'irja* is made of one disk revolving around a natural fixed point, the ascendant, which is conventionally represented at the nine o'clock position.

---

460. *Muqaddimah*, trans. Rosenthal, vol. 3, p. 172 [III, 138].

461. Humboldt, Systeme von Zahlzeichen, p. 216: "Im Orient wird *raml* die negromantische Kunst des Sandes genannt. Ganze oder gebrochene Linien und Puncte, welche die Elemente vorstellen, leiten den Weissager. [...] [D]ie notenartigen, knotigen, oft gebrochenen Parallellinien der Zauberbücher [...] scheinen nur graphische Projektionen von diesen Rechen- und Denkschnüren [, die den Abacus ausmachen]." The German word "negromantisch" is not found in any of the relevant dictionaries and probably constitutes a misprint; translation from the German, and emphasis added by D.L.

462. Lohr, Christianus arabicus, p. 64. Lohr mainly refers to the *Ars Compendiosa Inveniendi Veritatem* (1273–75), the *Ars Demonstrativa* (1275–81), the *Ars Inventiva* (1289–90) and the *Ars Generalis Ultima* (1308).

The Majorcan philosopher replaces the outer reference by a duplication of the artefact, which now only correlates to itself.

– On the *zā'irja*, truth is dynamically generated, by processing single signs. The same is true of the Majorcan device, which recombines letters in a much more simple and straightforward way. Here, the algorithm does not need to perform the seemingly impossible, to generate words, because the symbols are already thought to represent concepts.

– While the letters B to K indicate varying contents in the *Art*, like the qualities of God or the categories of thinking, the signs are continuously converted between the alphabets on the *zā'irja*. The same symbol may indicate a letter or a numeral and be translated between the two in different ways, depending upon which part of the system is employed.

The fact that only consonants are written down in Semitic languages permits the meaningful interpretation of many random permutations of symbols, which seems rather incredible when seen from the point of view of a vowel alphabet. The German linguist Johannes Lohmann compared the differentiation of Indo-European and Semitic languages to the separation of science (*epistème*) and poetry (*poìesis*) in Ancient Greece, the "thinking" and the "acting" components of sign usage. "In the Semitic word form," he wrote, "the semantic act is part of the living language."[463] This "magic" state of writing, which may have fostered the view that it was God-given, was illustrated in a more secular form by Claude Shannon in 1948: "The redundancy of a language is related to the existence of crossword puzzles. If the redundancy is zero, any sequence of letters is a reasonable text in the language and any two-dimensional array of letters forms a crossword puzzle."[464]

When Llull imported the *zā'irja* technology into the realm of the vowel alphabet, it lost its function altogether because of the higher redundancy within this type of writing system, in which many possible letter combinations are meaningless. In a notation in which many shufflings of the signs produce a meaningful word, their free permutation arises quite naturally and playfully as a symbolic *modus operandi*.[465]

---

463. Johannes Lohmann, *Philosophie und Sprachwissenschaft* (Berlin, 1965), p. 51f.

464. Claude E. Shannon, A mathematical theory of communication. *The Bell System Technical Journal* 27. 3–4 (1948): 379–423 and 623–656, reprint pp. 1–55, quotation p. 15. On the lower redundancy of Semitic writing see Sampson, *Writing Systems*, pp. 92–94.

465. In biblical Hebrew, רבע (RBA) means "to lie," רעב (RAB) "to starve," ברע (BRA) was the name of a king of Sodom, בער (BAR) "to put sth. away," ערב (ARB) "to exchange," עבר (ABR) "to overflow," while in English, the only two meaningful permutations of these three letters are "bar" and "bra"; cf. Wilhelm Gesenius, *Hebräisches und Aramäisches Handwörterbuch über das Alte Testament* (Berlin, 1962). The reason for this is that the Semitic writing system clings less closely to the mouth, cf. Chapter 2. Unpronounceable letter combinations are almost impossible in Arabic.

Scrambling T-R-U-T-H

More than 600 years ago, the *zā'irja* already provided a functionality which the constructors of search engines, and many others, are longing for today.[466] Knowledge of the technique appears to have been forgotten, or may have been applied to other fields, like in Brecht's parable on the obsolescence and the cyclical return of artefacts from his radio theory:

> "I recall an old story in which someone demonstrates to a Chinese man the superiority of Western culture. He asked: 'What do you have?' The answer: 'Railways, automobiles, telephones.' – 'I am sorry to have to tell you', the Chinese man responded politely, '*we* have forgotten those already.'"[467]

At least this man knows what has fallen into oblivion. An archaeology of algorithms and their artefacts tries to prevent a state of affairs where certain things suddenly appear to have never existed, which occurs after an alarmingly short time in the modern era.

The aims of artificial intelligence and *'ilm al-ḥurūf* (the science of letters) coincide: to provide a truthful answer to any question posed. But the knowledge of every fact in future history represents a tragic and highly paradoxical situation. It excludes human freedom, because it is based on the assumption that the fate of everything has already been written down. Even though the actor knows beforehand what will happen to him for the rest of his life, he is unable to change it, not even the slightest gesture, and more cruel than in "Groundhog Day": a man suddenly exists in a different system of time, in which the same day is replayed throughout eternity.[468] Paradoxically, he is the only person capable of acting differently. The irony of the film is that, whatever he does, the end result will be the same. He will wake up the next morning to the same old day replayed. If, on the other hand, the question whether the destined future can be changed is affirmed, the Book of Truth must have been outdated when it was consulted.[469] In this case, it would not determine fate from the start, but only contain the calculated results of all prior action, like Laplace's "demon":

> "We ought then to consider the present state of the universe as the effect of its previous state and as the cause of that which is to follow. An intellect that, at a given instant, could comprehend all the forces by which nature is animated and the respective situation of the beings that make it up, if moreover it were vast enough to submit

---

466. In reverse, it has been suspected that "artificial intelligence" represented "Cargo Cult" – the senseless and archaic combination of technical objects not understood; cf. Richard Feynman, Cargo cult science. *Engineering and Science* 37. 7 (1974): 10–13; Kenneth Mark Colby, Modeling a paranoid mind. *The Behavioural and Brain Sciences* 4 (1981): 515–560, esp. p. 534.

467. *Brecht on Film and Radio*, ed. Marc Silberman (London, 2000), p. 37.

468. Danny Rubin and Harold Ramis, *Groundhog Day* (1993).

469. Action movies from the 1990s, like "The Terminator" (James Cameron, 1984), "Total Recall" (Paul Verhoeven, 1990), "Terminator 2: Judgement Day" (James Cameron, 1991) and many others, playfully circle around this paradox.

these data to analysis, would encompass in the same formula the movements of the greatest bodies of the universe and those of the lightest atoms. For such an intelligence nothing would be uncertain, and the future, like the past, would be open to its eyes."[470]

If the truth-telling device is constructed algorithmically, somehow with all events up to this point at its disposal, the paradox is seemingly resolved. In fact it suffices to pose hypothetical questions to it to re-introduce the contradiction, having decided beforehand to undertake every measure to escape the predicted fate: "Will the stone reach the floor if I drop it?".[471] It is easier to imagine that by some mysterious power the oracle created some anomalies in reality than having it literally force the hand of the human subject.

The *zā'irja* is based upon a similar assumption. *'Ilm al-ḥurūf*, the science of letters, belongs to the broader discipline of *sīmiyā* ("letter magic").[472] The term is usually derived from Greek σημεῖον ("sign" or "signal"), the words "semantics" and "semiology" stem from. However, an Arabic dictionary, the Muḥīṭ al-Muḥīṭ of 1870, suggests the etymology derives from Hebrew שם יה – "shem yah" ("name of God"), and the names of God "certainly play a large part in sīmiyā."[473] These practices appear to be based on the idea that all things were generated by permuting those 99 sign sequences, and that all future events are determined by their further unfolding, which proceeds perfectly regularly and according to laws. If the rules that guide this symbolic development can be determined, the course of the Real can be predicted and, consequently, the Truth revealed.

The many preceding pages explaining the *zā'irja* operation combined with the condensed expression in Figure 24 allow its high complexity to be estimated. The Russian mathematician Andrey Kolmogorov described the concept as follows:

"One often has to work with very long sequences of symbols. Some of them, for example, sequences of digits in a five-place table of logarithms, admit a simple logical definition and can correspondingly be obtained by computation [...] according to a simple program. Others, however, presumably do not admit any sufficiently simple 'regular' construction: for example, a sufficiently long segment of a 'table of random numbers.'"[474]

---

470. Pierre-Simon Laplace, *Philosophical Essay on Probabilities* [1825], trans. Andrew I. Dale (New York, 1995), p. 6.

471. If the decision was not taken beforehand, the demon would not dispose of all information that will determine the outcome.

472. *Muqaddimah*, trans. Rosenthal, vol. 3, p. 171f. [III, 137f.].

473. *E.J. Brill's First Encyclopedia of Islam, 1913–1936*, ed. M.Th. Houtsma, vol 7: S–Ṭaiba (Leyden, 1987), p. 425f. Also note the similarity of *sīmiyā*, the art of mixing verbal elements, and *ḥīmiyā*, alchemy (the processing of material ones), which immediately follows in Ibn Khaldūn's account.

474. Andrey N. Kolmogorov, Report on 24 April 1963 to the Probability Theory Section of the Moscow Mathematical Society, quoted in: A.N. Shiryaev, Andreǐ Nikolaevich Kolmogorov. A biographical sketch of his life and creative paths, in: *Kolmogorov in Perspective* (Providence, RI, 2000), pp. 1–89, p. 64; cf. A.N. Kolmogorov, Three approaches to the quantitative definition of information. *Problems of Information Transmission* 1. 1 (1965): 1–7.

In its high complexity, the procedure resembles an encryption algorithm. Letters are converted to numbers, looked up in different substitution alphabets, and finally transposed. At the time of Ibn Khaldūn's initiation (1370), cryptography was already advanced among the Arabs, as described in detail by Ibrahim Al-Kadi. Among the reasons for this that Al-Kadi furnishes are: the translation of treatises from dead languages, some of which, dealing with alchemy, were enciphered; linguistic studies of Arabic; and the need to protect sensitive information in the administration of the state.[475] In the thirteenth century, the scholar Ibrāhim Ibn Moḥammad Ibn Dunainīr (1187–1229) devised an arithmetical cipher whose procedures are remarkably close to those of the *zā'irja*. First, the single letters of the cleartext are converted to numerals according to the *abjad* system, and then they are decomposed into two summands, which are translated back into alphabetic signs. In this way, "ḥ" (ح) is transformed to 8, and then to 2 + 6, resulting in "bo," بو, or to 1 + 7, "az" (از).[476] Alternatively, the value is simply doubled, 16, and expressed as a composite *abjad* numeral, 10 + 6, يو, which can be read as "yo," or tripled, 24, ىد, "kd."[477]

In 1467, the Italian polymath Leon Battista Alberti utilised the concentric circles that Llull had used to meditate and prove the truths of Christianity, that is, still in a revelational manner, to perform the opposite task: the concealment of communication. Previously, the outcome of wars had been predicted by scrambling letters on the *zā'irja*; now messages containing the future of the battle, in the form of commands to be executed by the units in the field, were made unreadable for the opponent by employing the same techniques. The aim was to remove the regularities of language, which would have permitted the strategic communication to be read, and no longer to imitate the semantic laws of Nature's unfolding to predict events to come. Moreover, in order to be decodable by the receiver, this operation needed to follow strict rules, whose traces also had to be concealed. Remarkably, in 1933 this paradoxical situation made it possible for the young Polish mathematician Marian Rejewski to decipher the first machine of this kind employed on a large scale, the German *Enigma*.[478]

475. Ibrahim A. Al-Kadi, Origins of cryptology: The Arab contributions. *Cryptologia* 16. 2 (1992): 97–126, esp. pp. 98–101. On p. 103, he lists Arabic cryptologists and their publications from the eighth to the fifteenth century.

476. The letter "w" also indicates the vowel "o" or "u" in Arabic.

477. Al-Kadi, Origins of cryptology, pp. 115 and 119, Fig. 7.

478. Władysław Kozaczuk, *Enigma. How the German Machine Cipher Was Broken, and How It Was Read by the Allies in World War Two*, trans. Christopher Kasparek (Frederick, MD, 1984). The combination of the different cyclical components of the *zā'irja* is in fact quite similar to Enigma. Cf. Chapter 5.

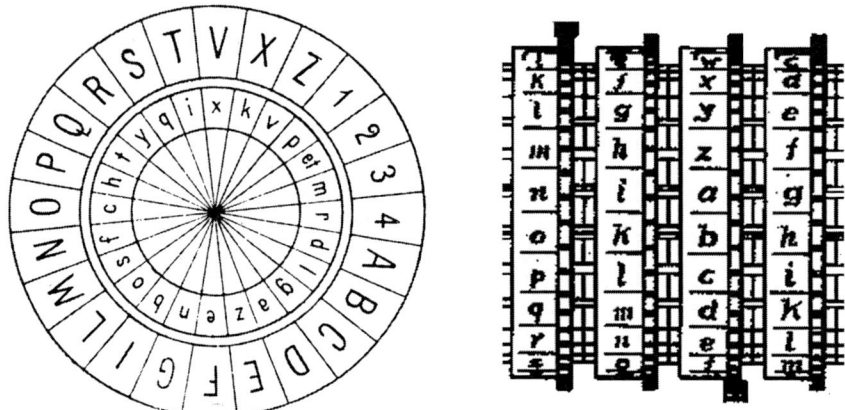

Fig. 25: The career of a technical form: Alphabetic disk of Alberti, 1467, and letter wheels within Enigma.[479]

The *zā'irja's* Effectiveness

The question whether the *zā'irja* was able to fulfil its function leads into the field of the Undecidable. To predict the behaviour and output of an algorithm in a formal way, for example, whether or not it stops at all, has been demonstrated to be impossible in many cases.[480] What can be said is that the complexity of the routines employed, substitution, transposition, and some simple arithmetical operations, are not alone sufficient to generate intelligible language, not to mention regular rhythm and rhyme. It all depends on the quality of the basis of the procedure, the seemingly arbitrary, rule-less two tables and the poem. If they were constructed following a simple regularity, it could be proven that the device cannot perform the claimed function. The late remnant of the *zā'irja* procedure that the British Orientalist Edward W. Lane (1801–1876) found in Egypt and published in 1836 depends entirely on the craftsmanship of the lookup table.[481] Starting from a letter that he selected blindly, the soothsayer picked every fifth, resulting in one of the hardcoded recommendations, like "Do it without fear of ill." The operation is similar to the final columnar transposition executed on the *zā'irja*, and the device is still ascribed to Idrīs.

479. Kahn, *Codebreakers*, p. 128; Arthur Scherbius, Ciphering machine (Patent no. US 1,657,411, filed in Germany as no. DE 383,594 on 12 February 1922).

480. Alan Turing, On computable numbers, with an application to the Entscheidungsproblem. *Proceedings of the London Mathematical Society (Ser. 2)* 42 (1937): 230–265. Any non-mechanisable investigation, on the other hand, usually fails at the complexity. Turing devised "oracles" capable of answering unsolvable questions; cf. A. Turing, Systems of logic based on ordinals. *Proceedings of the London Mathematical Society (Ser. 2)* 45 (1939): 161–228.

481. E.W. Lane, *An Account of the Manners and Customs of the Modern Egyptians*, 2 vols. (London, 1836), vol. 1, p. 336ff.

d	w	w	a	w	o	h	a	b	h
i	o	i	s	o	t	d	t	t	w
w	o	a	a	a	i	e	n	i	i
t	s	d	n	t	h	i	a	u	e
o	t	t	n	t	u	w	t	d	h
t	i	a	e	s	f	l	i	n	u
e	l	n	j	c	a	d	t	o	c
r	o	h	y	e	o	w	y	p	e
f	r	w	e	d	i	o	i	a	e
l	n	s	c	t	l	g	h	e	h

Fig. 26: Late Egyptian zā'irja.

In the original *zā'irja*, the initial question only enters the procedure as the number of its letters, and this happens towards the end, in the so-called "results," when most of the signs, apart from the last seven, have been determined. By contrast, two values are devoted to the world, the objective stellar situation. At the beginning of the passage, Ibn Khaldūn wrote:

"A question may have three hundred and sixty answers, according to the degrees (of the firmament). The answers to one question under a given ascendant differ in accordance with different questions (forming part of the question asked), which are referred to the letters of the chords of the *zā'irajah* and (in accordance with) the operation applied to finding out the letters from the verse of the poem."

The signs of the question are added to the ones from the chords, and their number counted. The result has to lie between 88 and 96, otherwise the text entered needs to be shortened.[482] Consequently, the input to the system consists in 30 (degrees) × 12 (possible rulers) × 9 (different numbers of letters) = 3,240 discernable "situations" or possibilities to which it generates an answer. It is difficult to say if the routine depended on the question in other ways. An enigmatic remark in the sixth cycle, after the first half of it has been executed, seems to imply that:

"At this point, one looks at the letters of the question. The (letters) that have come out (in the preceding operation) are paired with the verse of the poem, beginning at the end. One marks them with the letters of the question, so that it enters numerically into the verse of the poem. The same is done with every letter that comes out hereafter, in correspondence with the letters of the question. All letters coming out are paired with the verse of the poem, beginning at the end, and a mark is put on them."

If the only data that enters the routine is the number of signs, it is possible that each time it was consulted the astonishingly complex procedure only produced one of approximately 3,000 answers, which were in some way hardcoded in the tables and the poem. If no mechanisms exist that change the flow of the algorithm, it always generates the same text for a certain set of input values. In this case it would be functionally identical to its late Egyptian successor, only much larger. It is difficult to believe that such vast complexity would be employed to

---

482. *Muqaddimah*, trans. Rosenthal, vol. 3, p. 196f. [III, 162].

merely select one out of a set of different answers. A promising line of enquiry might be a frequency count of the tables, to find out how close they are to the normal distribution of letters in Arabic, as well as further cryptanalytic processing.[483] Another possibility would be to question the device directly, for example, by asking:

هل زايرجة تظهر الحقيقة؟ – "Does the *zā'irja* show the truth?"

The artefact can be regarded as a very early experiment in the free algorithmic processing and conversion of symbols. The operations are executed like calculations, but follow rules different from the mathematical ones, being derived from another kind of truth. Signs are freely transformed, guided by signs. If Ibn Khaldūn really was instructed in the technique in around an hour, then the level of penetration his text shows is astonishing. However, it seems that some of the information escaped him. Further reconstruction of the complete routine may be possible using the numerous other manuscripts on the subject written by authors who have since fallen into obscurity. It is hoped that the foregoing detailed analysis provides a more solid fundament for such an undertaking. "And God knows better."[484]

483. Al-Kadi has published the comparison of a frequency count he conducted with one that was performed more than one thousand years earlier by Al-Kindi; cf. Al-Kadi, Origins of cryptology, p. 112, Fig. 4.
484. *Muqaddimah*, trans. Rosenthal, vol. 3, p. 171 [III, 137].

Programming ENTER
Christopher Strachey's Draughts Programme

This chapter details some problems – and some solutions – encountered when resurrecting a programme for the game of draughts from 1951 on an emulator of the Ferranti Mark I. This machine was the industrial version of the Manchester Mark I, whose prototype, the Manchester "Baby" (SSEM), performed its first calculation on 21 June 1948. The algorithm described here was one of the earliest complex applications authored on the pioneer computer that served purposes beyond system testing. Christopher Strachey, an outsider to the Manchester computer laboratory and a schoolteacher, developed the software in his spare time.

The material relating to the draughts programme has been preserved in the Strachey papers in the Bodleian library, Oxford.[485] In it, there are approximately five versions of an algorithm that is about twenty pages long, pencilled on the standard Manchester coding sheets. There are also printouts of sample games Strachey played against the machine at the time, which were recorded on the teleprinter. Dates on the papers indicate the software was mainly developed in June and July of 1952. A first, undated version was probably written prior to May 1951.[486] In February 1951, the Ferranti Mark I was installed in Manchester University. Strachey gave a lecture about Draughts at the ACM conference in Toronto in September 1952, which was published in the Proceedings.[487]

---

485. Bodleian Library, Oxford, Special Collections and Western Manuscripts: Papers and correspondence of Christopher Strachey (1916–1975), CSAC 71.1.80/C.27–C.33. All the manuscript material referenced below is found in here.

486. In a letter dated 15 May 1951, Strachey wrote to Turing: "I have completed my first effort at the Draughts" and he was obviously talking about the Manchester Mark I. At this point, the algorithm already had "input and output arrangements." Cf. Strachey Papers, CSAC 71.1.80/C.22: Strachey's copy of his letter to A.M. Turing, 15 May 1951.

487. Christopher Strachey, Logical or non-mathematical programmes. *Proceedings of the ACM* 1 (1952): 46–49. Cf. Alan M. Turing, C. Strachey, M. Audrey Bates and Bertram V. Bowden, Digital computers applied to games, in: *Faster Than Thought*, ed. B.V. Bowden (London, 1953), pp. 286–310.

Game Machine User Experience

When the software started, it asked the user to PLEASE READ THE INSTRUCTION CARD on the teleprinter. He would then hit a key labelled "KAC" on the console to signal he had done so. The algorithm asked him to spin a coin and claim either heads or tails. The user let the programme know via a switch and KAC if it had won or not to determine who had the right to start the game. Then human and machine made moves alternately, the latter by printing them on the teletype, the former by setting the hand-switches on the console and hitting KAC. The complete game was printed out, and two consecutive situations could always be visually compared side by side on cathode ray tubes 3 and 5, which were part of the working memory of the machine (cf. Figure 1). This software presumably constitutes the first usage of a graphic display in a computer programme.

Fig. 1: The draughts board as shown by the storage CRT of the Ferranti Mark I, and a modern recreation by the author.

Strachey had coded an additional "preview feature": After the user had announced his move by setting it up on the switches, the machine showed the resulting position on cathode ray tube 3. If he then answered NO by composing ///T on the console (bit 19 on), the algorithm reverted to the previous situation on the board and he could try something else. If the user input wrong information, the machine became increasingly angry, until it uttered: I REFUSE TO WASTE ANY MORE TIME. GO AND PLAY WITH A HUMAN BEING/.[488] A similar routine existed if the opponent took too long to reply. Strachey had apparently become fascinated with the slightly obscene theatrical effect of a machine making human-like statements and showing "emotion." His next software was an inversion of this rather strict, impatient character, a programme for the composition of love letters. Draughts already contains the complete "rhetoric" that is needed for it algorithmically, including the selection of pre-fabricated text based on random numbers.[489]

---

488. The slash was probably used as an exclamation mark, which was missing in the teleprinter code.
489. Cf. Chapter 3.

Coding a Game

For the coding of the situation on the board, the white fields had been numbered from 0 to 31, and three 32-bit variables (memory locations) named B, W, and K respectively expressed the positions of black pieces, white pieces and kings of both colours, by setting the corresponding bit = 1 (cf. Figure 2).

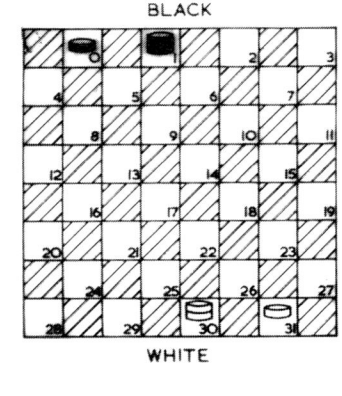

BLACK

WHITE

B = I Ioo oooo oooo oooo oooo oooo oooo oooo

W = oooo oooo oooo oooo oooo oooo oooo ooI I

K = oIoo oooo oooo oooo oooo oooo oooo ooIo

Fig. 2: Coding of positions in Draughts.

On the other hand, a move sequence consisted of two values in the same range, the fields from which – and to which – the piece was displaced, whereupon the position of a captured piece could follow, for example 23–14, with the opponent's piece hit on field 18. The programme also mastered multiple captures correctly. For setting up moves on the hand switches, Strachey employed an intuitive system rather close to decimal, where the first five bits indicated the tens (0 to 3), and the second and third the units (0–9) of the position number. In this way, sequences like "23–14 (18)" could be expressed as:

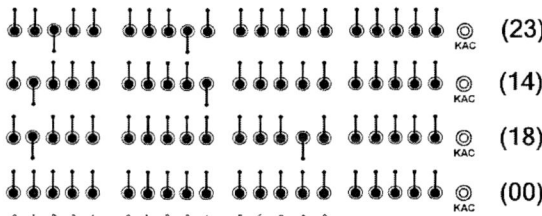

(23)

(14)

(18)

(00)

0 1 2 3 4   0 1 2 3 4   5 6 7 8 9

Fig. 3: Coding of move sequences.

The first bit in each group signifies 0, the second 1, and so forth. To end the move sequence and return control to the machine, the user had to hit KAC with nothing set.

The strategy implemented in the game algorithm was a heuristic one, so one could also claim draughts was the first heuristic programme. Strachey wrote that the difficulty of "the machine to look ahead for a number of moves and choose its move by a valuation scheme" was of great theoretical interest and presented a typical example of a "large logical programme." Other games were less challenging, because "most of them use the complete mathematical theory so that

177

the outcome [...] is no longer uncertain."[490] His programme calculated the next move by looking ahead for a certain number of steps without having an overview of the complete game. By not trying to exhaust the endless number of combinatorial possibilities, he "succeeded in making a programme [...], which will play a complete game of Draughts at a reasonable speed." In fact, this is not true: There is no code to control the end game, to detect it is over and to announce a winner. To write a programme that could handle the rather complex task of playing draughts must have been sensational at the time.

The central element in the heuristics of the algorithm was the evaluation function for future positions. In it, the machine calculated all possible moves up to a certain depth and summed up the material left on the board resulting from each, counting three for a king and one for a normal piece. Theoretically, i.e., from the perspective of storage space, the algorithm could look ahead three operations on each side (with depth = 6), but in fact, due to the much more pressing limits on time, it was in most cases only anticipating three in total (depth = 3). Actual programme performance can be very different from the planned one, and Strachey's strategy had the serious flaw of causing the machine to behave suicidally: As a result of the valuation scheme, it sacrificed all its pieces when the opponent was about to king. Strachey met this by adopting two different depths of search in such a way that as long as one of the last two moves had been a capture, the machine calculated on. After that, it kept looking ahead until the second depth value was reached.[491]

Strachey had separated the strategic core of the algorithm from the service functions and commented: "It is rather typical of logical programme: that this organising routine is in fact longer than the game-playing routine proper." The latter was called DRAUGHTSB or DR/B and consisted of eight pages (in the version dated 10. 7. 1952), while the service part (DRAUGHTSC) occupied another ten sheets, with four containing auxiliary functions. So the programme comprised 18 or 22 pages in total, depending on the method of counting – incredibly long for the time.

---

490. Strachey, Logical or non-mathematical programmes, p. 47.
491. In the run on 10. 7. 1952, this value (*b*) had been 1, with *a* (normal search) = 5.

Resurrecting Draughts

In the course of software reconstruction, usually some parts start to work, while others still malfunction and ultimately lead to a crash of the programme one tries to resurrect. One technique here is to follow the algorithm along to find the exact point where it starts to go wrong. This is usually earlier in the executed code, slightly before the final crash. (It is astonishing how long programmes can sometimes run on completely wrong grounds.)

When the exact position of the aberration is found, this particular place in the code can be investigated and probably fixed, provided the situation is not too complex. The software will then continue to execute, until it encounters another crash point, or ideally run through to the end, in which case the reconstruction succeeded. This technique of debugging already existed in the 1950s and there were dedicated "check sheets" to trace or log a programme at runtime, i.e., to record the memory locations that changed in the sequence of the operations of the algorithm.

In one such situation in the beginning of the resurrection of Draughts, the programme was waiting for some time, and then went to a "hoot stop." This was the symbolic equivalent of a crash, by which the software signalled that something had gone fundamentally wrong.

Upon closer inspection, the algorithm was stuck in the following lines[492] (see the note at the end of this chapter for an explanation of the notation):

```
1 - T/: Accumulator (A) = 147456
2 - TN: A = A-1
3 - /M: go to line 2 if A >= 0
4 - /I: switch M and L, the left and the right side of (A)
5 - /H: continue execution of programme if A >= 0
6 - /T: go to hoot stop
```

In line 1, the 80-bit accumulator is set to a rather high number, 147456, by copying it from address VK in the working memory. It then decreases this quantity by subtracting the contents of address E: from it, which holds 1. This location is part of two pages of values that are permanently kept in memory, called PERM.[493] The third line is a conditional statement: If the accumulator is greater than 0, go to operation 2, where the number is again decremented. At one point, the value there will change from //////// //////// to ££££££££ ££££££££, that is, from 0 to -1. Since a command takes 1.2 milliseconds to complete on average, this will happen after approximately 5.9 minutes. The algorithm then continues in line 4. The operation here exchanges the left (L) and the right (M) 40 bits of the accumulator. Since it is set to all 1s, this produces

---

492. C. Strachey, Draughts programme, DR/C 1, lines T/ff.

493. Cf. R. Anthony Brooker, ed., *Programmers' Handbook (2nd Edition) for the Manchester Electronic Computer Mark II* (typescript, Manchester University, October 1952), p. 3.23: PERM, line NSf.

the same number, ££££££££ ££££££££, which is -1. In line 5, the algorithm jumps to what is obviously the continuation of the programme, if and only if the quantity in the accumulator is positive! Otherwise it enters the already-mentioned hoot stop – an endless loop with no break condition, which consists of the following two lines:

```
1 - /V: hoot
2 - /P: go to 1
```

In modern notation, the algorithm we just discussed could be rewritten in the following way:

```
int i = 147456;
while(i >= 0) i--;
switchMandL(i);
if(i < 0) hootStop();
else continue programme execution
```

This code seemed to make no sense at all! To understand it, it is useful to consider how signed numbers were represented in the Manchester Mark I. Generally, these were 40 or 80 "binary digits," written with the most significant bit to the right. The handbook specified: "On the plus-minus convention the most significant digit is used to represent the sign."[494] To find out if the number in the 80-bit accumulator was positive, it was sufficient to have a look at bit [79]: When it was 1, the number was negative. The machine automatically copied the value of this bit to the A-sign flip-flop, and when it encountered an A-conditional statement, it referenced the data there. So again: How could the switching of the two sides of an accumulator full of 1s result in the 79th bit becoming zero? Apparently, the algorithm expected something that could never happen, an impossible event. In other words, it was waiting for a miracle.[495]

In the operating instructions, Strachey wrote that the "machine gives a 'pip-pip' signal when it requires attention. It should always be restarted by operating KAC after it [the machine] has been set appropriately." He went on to give examples of what the computer could say and different ways of responding to it. The KAC key was one of the several clearing switches the Manchester Mark I inherited from the "Baby" prototype and its function was to empty the accumulator. But would hitting KAC not lead to the same situation as counting it down until it reached zero? In both cases, the accumulator would first become all zeroes, = 0, and then all ones, = -1, when it was decremented in line 2. It was impossible to see any reason why the switching of the two parts would make the number positive. And yet, it was quite obvious that the code in question could do exactly this: tell the difference between counting down and hitting KAC.

---

494. Brooker, Handbook, p. 1.8f.
495. In very much the same way, the tautology while(true), which encloses the run loop in the core of most programmes, can only be broken in the improbable event that truth is no longer truth. Cf. Chapter 1.

Analogies in Logical Design

The solution to the riddle was that Strachey's programming relied on the logical design of the Mark I, namely its hardware properties. I failed to make sense out of the code fragment for a rather long time, because I was looking at the machine on a purely symbolic level, where signs were transformed into other signs instructed by signs. The emulator was only an implementation of the Mark I's operation codes and its effects on the contents of the stores. In this mode of thought, pushing a button was treated like a command, and more importantly, like a synchronous one. There was no difference between KAC and the operation code T:, which also cleared the accumulator.

In writing, meaning is conveyed by material elements, the words and letters. In the same way, the data and operations in computers are represented by certain real systems with suitable properties, by a physical analogy. The function to clear the accumulator is implemented in certain electronic components, a Williams tube by the name of "A," in a way that follows the logic of this device. Since something is stored here only if it is refreshed, preventing it from recirculating is equivalent to deleting the data.

But it is not only the spatial or physical analogy that counts, but also the temporal aspects of this simulation of thought processes. On the most basic level, computers move in cycles, which are subdivided in a number of phases in which certain predefined elementary actions take place. In the Manchester Mark I, there were seven of them: SCAN1 to SCAN3, and ACTION1 to ACTION4. The timing with respect to these also determined if an operation was synchronous or asynchronous. In the logical design picture of the machine we are interested in here, it is important to know in which phase certain parts of commands are executed.

Programming ENTER

What different activities are happening on the logical design level in the algorithm while it counts down the accumulator from 147456 when compared to its behaviour in response to the user hitting the KAC key? The eighty bits of A are set to 0 and the software subtracts 1 from it, making it negative, the A-sign flip flop is set and the programme breaks from the first loop. What is important here is that the sign of A is identified after the arithmetic, but before the number re-circulated returns to the accumulator. Upon leaving the loop, the number (-1) is not re-circulated and hence A is empty again. When the algorithm switches M and L in a later cycle, the A-sign flip flop is clear and the programme jumps to its continuation, not to the loop stop. The rather elaborate sequence thus simply detects if the KAC key has been pressed. In that case, the software jumps into the following code fragment:

```
1 - /J: M += //// ///E
2 - /H: go to 1 if A >= 0
```

First, a number is added to the right part of the accumulator, equivalent to adding 1 into its 75th bit. Then, if the number is not negative, the procedure is repeated. Again, it seems quite impossible that by adding a positive quantity, the result can become negative. Obviously, the user is still holding down KAC when these statements are reached, which prevents recirculation, leaving the accumulator empty. Once the key is released, it raises in increments, until this carries over into the sign bit at the 16th addition. The algorithm jumps to its continuation. The code thus detects the release of the KAC key and waits if it remains stable in the "up" position, to prevent the accidental bouncing of contacts that might disrupt user interaction by simulating another key press. With the fragments described above, it formed a detection sequence for the typing (press/release) of the key.

"Phew! that was a good exercise," wrote Christopher P. Burton after he had found out what most of the mysterious fragments meant. So the code was actually not waiting for the impossible. Strachey had simply constructed in software what would today be called an ENTER key. It was needed to allow the user to communicate with the software: The operator prepared the console's hand-switches to play the next move and transmitted the information to the machine by depressing KAC. Interestingly, no ENTER key to "send" the carefully composed data to the machine existed on the "keyboard" of the Mark I. But luckily enough, with some ingenuity, it could be programmed.

I/O on the Mark I

If Strachey had to program an ENTER key, what switches and buttons were there and why was such a key missing? The programmer was operating what the Mark I prototype's control panel had evolved into – a console with many switches, into which so-called typewriter buttons were integrated. Not much had been added since the prototype's first appearance in June 1948. There were switches to set and remove single bits in store, switches to compose line numbers and function numbers manually, clearing keys for the different stores of the machine (one of them being KAC), switches for selecting what the monitor tube was showing and keys to control the stepping of the machine. Apart from some switches dealing with I/O equipment, there was only one row of keys on the console of the Ferranti Mark I that was new: a row of twenty so-called "hand switches" that could be freely programmed in the sense that their value could be copied to a memory location and manipulated at will.

At the time, the idea of how the user would interact with the machine was that he would set up certain parameters and possibly input values on the switches, then start the machine and wait until it would come to a halt. If this was foreseen, additional values could be entered before continuing the programme by operating the L or G-switch. The command to copy the hand switches would then be the first after the stop command. The way in which Strachey describes user interaction in Draughts can be regarded as paradigmatic for the time: The machine "should always be restarted by operating KAC after it [the machine] has been set appropriately."[496] Possible input a user could have while the software was running had not been planned for, and this is why the Mark I only employed the teleprinter's capabilities to print out, and not to input data via a "keyboard" and in this way "talk" to the machine. This would have required a buffer to hold the key codes received. The method on the Mark I to stop the machine altogether, compose information on the switches, and restart, stands at the beginning of a history of solutions to the complex problem of organising the flow of information to and from the human user. Although the operator from the times of Morse code had been removed, there still needed to be something that was listening at the moment of the keystroke to allow for communication.

The KAC key was not supposed to be hit while the program was running, but before: to clear the different stores of the machine. By trying out a combination of hardware elements and user interaction that had not previously been envisaged, Strachey stumbled upon some new, yet undefined behaviour of the machine: If KAC is pressed while the accumulator is counting down, it nevertheless always remain positive.

The History of ENTER

From a computer-historical perspective, the teleprinter was already connected to the computer early on. Apart from the Manchester Mark I, there was also the US-American Binac in 1949, where it was used to input data onto magnetic tape and output onto paper. Until the late 1960s, the artefact became increasingly integrated with early calculating machinery, constituting the first computer terminals to share the time of mainframe computers. The Teletype Model 33 ASR, introduced in 1963, was the most successful paper teleprinter used in this function. With the growing communication of data between user and machine, the teleprinter started to appear slow, and in 1967, the first "glass teletype" was introduced, the Datapoint 3300, which emulated the Model 33. Instead of wasting paper on something as ephemeral and profane as the communication to and from a computer, the data was now stored in a volatile form of visual memory on a CRT: 25 rows of 72 columns of upper-case characters. It supported control codes to move the cursor up, down, left and right, to the top left of the screen, or to the start of the bottom line.

---

496. Strachey, Draughts Instructions.

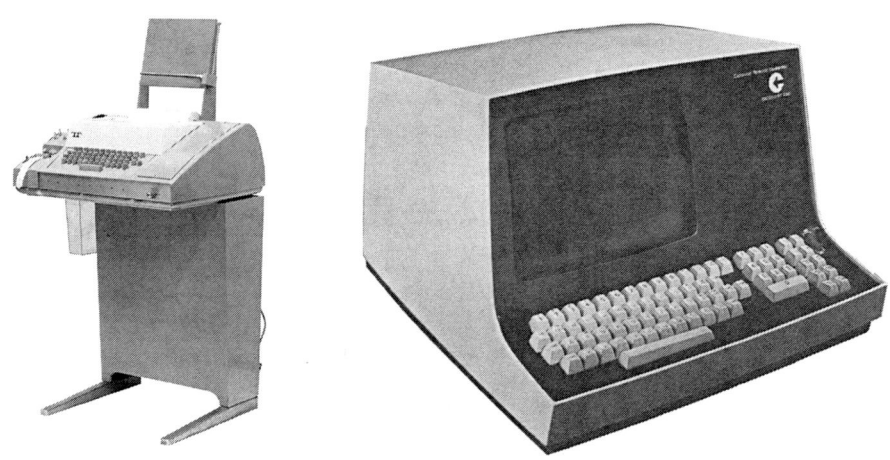

Fig. 4: Teletype Model 33 ASR, 1963; Datapoint 3300, 1967.

With the infrastructure of mainframe computers and user terminals and the advent of time-sharing (which was actually patented by Strachey in 1959), the concept of a proper ENTER key also evolved – on the keyboard of the IBM 3270 series terminal, 1972:

> "The terminal would display a full page of information. Either 80 or 132 characters wide and 24 or 27 lines long. This information was held in memory. The unit could be formatted as an entry form so there would be places on the screen to fill in the information. You could tab or new line (this key is where the return key is normally located) from field to field. When you had everything entered you would press the enter key (to the right of the space bar) and only the entry fields would be sent to the mainframe. This minimised the amount of data transmitted and as it was sent in bursts several terminals could use the same port into the mainframe."[497]

When the "glass terminal" replaced the paper of the teleprinter, its buttons transformed into the keyboard we know today, and this constitutes the only material remains of the communication device. The teleprinter always consisted in two parts, the sender/keyboard and the receiver/printer, which could be combined in a free way, forming a typewriter as a special case of a more general class of machines only if the output of the sender (a contact named B) was connected to the input of the receiver on the same machine (W2). Finally, twenty years after Strachey had programmed ENTER, the function key started to exist on what remained from the tele-typewriter, the computer keyboard.

---

497. David Christ, US-American radio amateur, personal communication, 3. 8. 2012.

Fig. 5: IBM 3270 series terminal, 1972; keyboard detail.

Not only did the teleprinter survive materially in the form of the keyboard, but also immaterially, as a mental form: The first protocol ever created was the one used on the teleprinter, which became the prototype for all serial communication: TTY. It was patented by the US-American inventor of the teletype, Howard L. Krum, in 1910:

> "[T]he signals or characters are distinguished by different permutations of a definite number of impulses of opposite polarity. […] [T]here are thirty-two permutations of positive and negative impulses when there are five impulses in each permutation, so that thirty-two different signals can be transmitted. […] [I]n order to obtain the desired thirty-two different combinations or permutations, it is necessary to transmit a starting impulse which is not varied but is always of the same polarity."[498]

The definition of the start bit is what makes the momentous difference between the formulation of a protocol and the conception of a code, which had existed long before: the Baudot or Murray five-bit code. Whereas a code is the formalised representation of information, a protocol defines the exchange of information in time. The necessary start bit was the clock – it kept the two machines synchronised, the main problem in the early transmission of writing. After the start bit, which was always low because the line was high, there followed five data bits in this code, then a stop bit. The communication was bit-synchronous, but word-asynchronous, meaning that the number of bits in the word and their timing was clearly defined, while the time between two words was undefined. The only thing added to the original idea by posterity was one bit to the data part of the word, as a very simple means of error detection, indicating if the parity of the bits was even or odd. The very common designation "8N1,"

---

498. Howard L. Krum, Electric selective system (Patent no. US 1,286,351, applied for 31 May 1910).

which may be known to some readers from the era of modems, consequently means a TTY protocol with 8 data bits, no parity, and one stop bit. The RS232 port of the modern PC, to which the mouse used to be connected before USB was introduced, still talked very much in the same way as the teleprinters of times past.

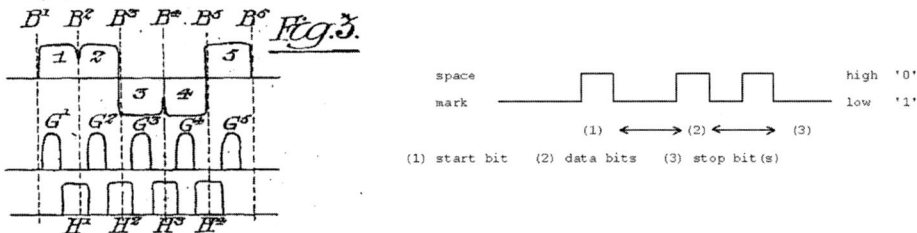

Fig. 6: Original schema of Krum's protocol, 1910, and modern schematic of "TTY."

The Limits of Simulation

The experience with Strachey's code for ENTER demonstrated that any simulation of machinery on the purely symbolic level was essentially incomplete. In addition to implementing the command set, it was necessary to emulate the logical design, the way in which current flowed through the computer to represent certain data and operations. The symbolic level I had implemented suddenly appeared to be the mere surface of what was really happening in the machine. And there were obviously other planes that mattered, for example the hardware properties of the components involved, which in their combination surely also created "strange" effects.

In fact, the strength of Turing's concept was that his universal symbolic machine was independent of any concrete implementation like delay lines or transistors. With such machinery, the ideal of Heideggerian *Zuhandenheit* is reached if the concrete hardware can safely be ignored. That it was in fact a lot of work to make the faultiness of the hardware disappear to reach this ideal realm of purely symbolic computation can be felt in documents like the engineers' log book of the Ferranti Mark I: "Prepulses disappear intermittently, but when the machine has obeyed order J:/N no known method of fingering the switches seems to get them back but some unknown cause will do so."[499] But even if the ideal purely symbolic state is reached, the hardware stays there as its Unconscious. It is always much more than what it implements. While certain properties of a system are in focus, because they represent the meaning, there are others that are unimportant in this respect and are ignored – they may return. The

---

499. National Archive for the History of Computing, Manchester, NAHC/MUC/2/C6: Mark I Log Books, 1951–58 (6 vols.); comment by Alec Glennie, ~17 January 1952.

Unconscious of the binary – which only knows on or off, no delitescence – is the slumbering-on of the hardware.

The experience of ENTER seemed to indicate that the only truthful resurrection of an old algorithmic artefact was in fact to reconstruct it from the original parts. A software resurrection can only implement what was specified at the time and not the machine that was actually built. I was very amused to learn that the "original parts" also behaved strangely at times. For the rebuild of the EDSAC, the Mark I's sister machine constructed by Maurice Wilkes and his team in Cambridge in 1949, the principal valve used, the CV1136, was no longer available in sufficient quantity. The rebuild team resorted to obtaining valves that had been manufactured decades later as replacements for customers who needed to maintain old equipment. Externally, they looked identical to the originals, retaining complete electrical and physical compatibility. But inside the metal can, a modern miniature valve was found. The hardware on the market also started to become simulative![500]

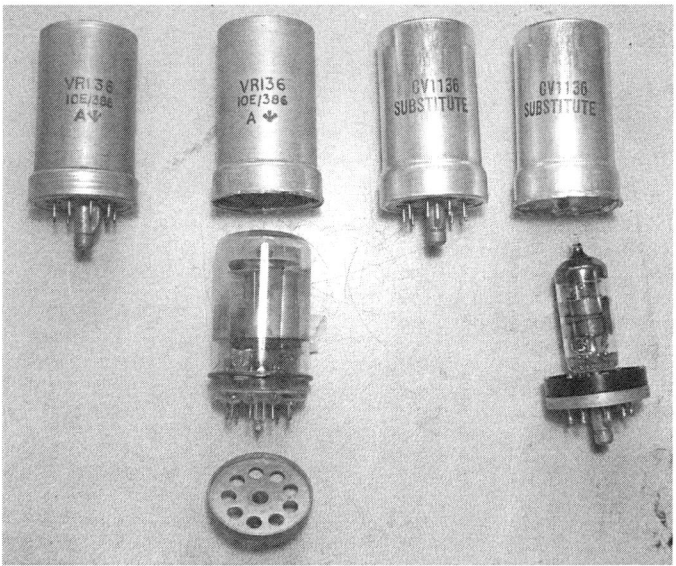

Fig. 7: Original valves and their replicas.

A pragmatic approach to this eventually came from Brian Russell, one of the demonstrators of the Manchester baby replica. He wrote the following when he heard I was considering an update for my emulator:

"Any model is a 'Simplification of Reality.' If your model runs most of the programmes that you want to run, then there is no need to make enhancements. If the object of your model is to better understand how the hardware works, then it might

---

500. Christopher P. Burton, personal communication, 31. 8. 2012.

be worth the effort of writing enhancements and it might be worth the increased run-time to model those enhancements. Whatever your model does it will not model everything. For instance, the original Baby was prone to malfunction in the presence of Electromagnetic Interference – it crashed when a tramcar passed along the tracks outside. You won't want to model that!"[501]

The Manchester Mark I and its Notation

The Draughts programme ran on the Ferranti version of the Manchester Mark I and Strachey used the notation established by Turing in the programming manual. The machine was based on a 20-bit word, and 20-bit numbers (and also instructions) were specified as four 5-bit elements, each element taking the name of the teleprinter code equivalent to the 5-bit value. Thus binary "00000" was expressed as "/" and "10000" (least significant on the left) as "E". The written form of numbers and instruction was quite opaque unless one was very familiar with all 32 of the possible teleprinter codes. The 5-bit value "11111" was written as "£".

Most instructions contained a function (operation) number and a store address. The function number was six bits long and could be expressed as two teleprinter characters, the first of which was always "/" or "T" ("00000" or "00001"). The instructions relevant to this article have the following meanings:

/H	Jump direct if accumulator >= 0	/T	Jump direct unconditionally
/I	Exchange most and least significant halves of accumulator	/V	Hoot (sound the loudspeaker)
/J	Add contents of a store location to most significant half of accumulator	T/	Load accumulator with contents of a store location
/M	Jump relatively if accumulator is >= 0	TN	Subtract contents of a store location from accumulator
/P	Jump relatively unconditionally	T:	Clear the accumulator

The Accumulator is 80 bits long, containing four 20-bit words. The most significant bit (bit 79) is the sign bit, 0 meaning that the number in the accumulator is zero or positive, and 1 meaning the number is negative.

---

501. Brian M. Russell, personal communication, 8. 5. 2012.

Programming Degree Zero
A Genealogy of Turing's Machines and Algorithms

In 1936, the 24-year-old Alan Turing theoretically conceptualised the computer in his seminal investigation "On Computable Numbers."[502] But as a pure mathematician, he only introduced the paper automaton of the Universal Turing Machine to show that even with such a theoretically perfect mechanism, it was not possible to compute the answer to all mathematical questions, a problem posed by the German mathematician David Hilbert at the turn of the twentieth century. In particular, he proved that the question of whether or not a certain algorithm would return with the answer in finite time was incomputable.

Although the concept was purely theoretical, there were certain technical developments that enabled such a machine to be imagined. Teleprinter networks communicated and handled the code of discrete signs in a very flexible way. The information was encoded in 5-bit Baudot and could be stored temporarily on 5-bit paper tape. Turing's tutor, Max Newman, wrote in 1955 about the Universal Machine: "It is difficult today to realise how bold an innovation it was to introduce talk about paper tapes and patterns punched in them, into discussions of the foundations of mathematics."[503]

Already in 1912, the London newspaper the "Daily Mail" adopted the teleprinter system developed by the Canadian inventor Frederick George Creed to transmit the paper from London to Manchester for same-day publication. In 1914, the Associated Press built up a private network of teletypewriters, as these were called in the USA, for delivering copy to the New York newspapers, and this spread quickly until more than 800 subscribers received their news dispatches by these machines. Delayed by the war, the Press Association in the UK did not set up a similar system until 1920, a private news network comprised of several

---

502. Alan M. Turing: On computable numbers, with an application to the Entscheidungsproblem. *Proceedings of the London Mathematical Society (Ser. 2)* 42 (1937): 230–265.

503. Max H. A. Newman, Alan Mathison Turing, 1912–1954. *Biographical Memoirs of Fellows of the Royal Society* 1 (1955): 253–263, p. 256.

hundred Creed teleprinters to serve practically every daily morning newspaper in the UK, and for many years this was the world's largest private teleprinter network. At the core of Creed's invention were two typewriters controlling each other electrically in a cross-over fashion via the first code – Baudot, and the first protocol – TTY.

Fig. 1: Russian caricature of a Turing machine; Creed 7B teleprinter; 5-bit paper tape with Baudot code.

The Zeta Function Machine and Lord Kelvin's Tide Predictor

It has been suggested that Turing, as a mathematician, did not really know how to deal with hardware. But according to Newman's obituary, Turing was interested in practical computing from at least March 1939 on, when he was 26 years old and wrote a paper on a method of calculation of the Riemann zeta-function. At the same time and with help from Donald McPhail, a Canadian engineer student at King's College, he created the blueprint for and started to build a machine at his home which would calculate "the Riemann zeta-function on the critical line for 1,450 < t < 6,000," an investigation up to 1,450 having already been undertaken by the British mathematician Edward C. Titchmarsh (1899–1963) in 1935/6.

In Turing's application to the Royal Society for a grant of £40 to construct the machine, he continued with as much self-confidence as accuracy with respect to practical relevance: "I cannot think of any applications that would not be connected with the zeta-function." At that time, the importance of primes had not yet been derived from the practice of cryptology, but from the special role they played for investigating the properties of natural numbers. Turing's other teacher, Godfrey Hardy, remarked about the pacifistic character of his field of work in 1940: "No one has yet discovered any warlike purpose to be served by the theory of numbers or relativity, and it seems unlikely that anyone will do so for many years." He would be proven wrong at least twice over the next few years.[504]

As a 15-year-old boy (around 1793), the German mathematician Carl Friedrich Gauss (1777–1855) had conjectured about the distribution of primes, which is dense in the beginning and then fades out, that "around a large

---

504. Andrew Hodges, *Alan Turing. The Enigma* (London, 1983), pp. 140f., 155–158.

number x, roughly 1 in every l_n x integers is prime," l_n x being the natural logarithm of the quantity. Only much later in 1859, the German mathematician Bernhard Riemann derived a more precise formula to estimate the density of primes at any given point by applying Euler's zeta function to complex numbers:[505]

$$\zeta(s) = \sum_{n=1}^{\infty} n^{-s} = \frac{1}{1^s} + \frac{1}{2^s} + \frac{1}{3^s} + \cdots \qquad \sigma = \Re(s) > 1.$$

It was the function of a complex variable s, i.e. a number with a real (x) and an imaginary (t) part which can be thought of as a point in a plane with coordinate (x, t). The zeroes of the zeta function, that is, the values of the variable s at which it vanishes, are particularly interesting for mathematicians, since they provide means to consider the distribution of primes. Some of these values ("trivial zeroes") can be immediately determined, but others ("nontrivial zeroes") cannot. Riemann conjectured, but did not prove, that for nontrivial zeroes the real part x always equalled 0.5, so that all of these values lay on the line x = 0.5 in the complex plane (x, t). This was referred to as the "critical" line. Turing's first machine should have calculated the real part x for the imaginary part t of this function in the interval 1,450 to 6,000 to find values off the critical line and in this way disprove the conjecture.

The design of his mechanical implementation of the formula was based on the tide-predicting machine that had been constructed in Liverpool by Edward Roberts in 1906, which was a direct successor of Lord Kelvin's (William Thomson's) original apparatus of 1872. A number of rotating wheels, whose amplitude and phase angle could be set, were connected together by a pulley system summing up the different forces. The machine was operated by a crank handle and calculated a year of tide heights and times for one location with a resolution of one hour within one and a half days.

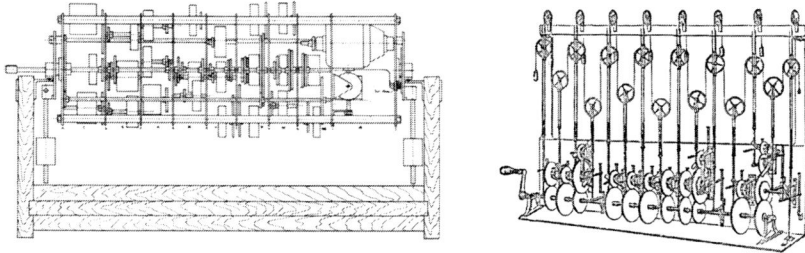

Fig. 2: Turing's zeta-function machine, July 1939, and Lord Kelvin's tide prediction apparatus, 1872.

505. Andrew R. Booker, Turing and the primes, in: *The Once and Future Turing*, eds. S. Barry Cooper and A. Hodges (forthcoming, online: http://www.maths.bris.ac.uk/~maarb/public/papers/turingprimes.pdf).

The Roberts' machine Turing copied calculated the tides for the D-day landings shortly thereafter. On 9 October 1943, the Admiralty's superintendent of tides at the Hydrographic Office, William Ian Farquharson, sent Arthur Thomas Doodson at the Liverpool Tidal Institute a three-page handwritten letter marked MOST URGENT. It included 11 pairs of harmonic constants for a location code-named "Position Z." He asked Doodson to produce hourly height predictions for four months commencing 1 April 1944. "The place is nameless and the constants inferred," he wrote. "There is in fact very little data for it. I am gambling on the inferred shallow-water constants giving something like the right answer." He was correct: The landings commenced on Tuesday, 6 June 1944, beginning at 6:30 am British Double Summer Time (GMT+2) when the tide was low.[506]

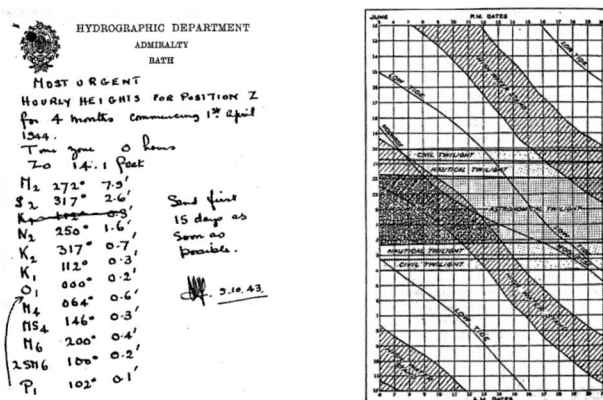

Fig. 3: Letter from William Ian Farquharson to Arthur Thomas Doodson, 9 October 1943; and output plot of tide predictions for D-day.

The zeta-function apparatus was a manual, continuous, analogue machine modelling a mathematical formula. It implemented the synthesis of thirty wave-like Fourier terms by employing about 80 gear wheels. The work on the propagation of heat by the French mathematician and physicist Joseph Fourier (1768–1830) enabled any equation to be expressed as a summation of wave functions. Fourier had conjectured, but not proven, that any function could be generated using his method.

Because the mathematical expressions were too difficult to calculate, the scientists assembled a brass apparatus from physical components like wheels and pulleys that in one aspect (for example weight or movement) approximated the desired mathematical function, and executed the calculations on this machine instead of in Arabic numerals. A complex formula that included terms derived from the geometric circle was translated by simple analogy into the physical behaviour of metal wheels and other objects from classical mechanics.

---

506. Bruce Parker, The tide predictions for D-Day. *Physics Today* 64 (2011): 35–40.

These elements, which had served a practical purpose, now implemented an operational, symbolic one.

In 1876, Lord Kelvin's brother, the engineer James Thomson, invented an integrating machine, the Tidal Harmonic Analyser, which is a predecessor to Vannevar Bush's Differential Analyser (1928–1931). This invention was the first semi-automatic waveform analyser and synthesizer. Lord Kelvin wrote in 1881: "The object of this machine [the Tidal Harmonic Analyser] is to substitute brass for brain in the great mechanical labour of calculating the elementary constituents of the whole tidal rise and fall, according to [...] harmonic analysis."[507] Babbage's Analytical Engine (1837) would have been an even more versatile brass apparatus, but it was never built.

Enigma and the Turing Bombe

Unfortunately for the discipline of pure mathematics, Turing had to abandon the calculating machine before it was finished when he was approached by the British government to help with some work of a more applied nature: to analyse and break into the cryptosystems of Germany and other nations. Since Leibniz, cryptography had been an integral part of *ars combinatoria* and is one of the main strands of this discipline that survives today, along with coding theory and group theory.

With the outbreak of war on 3 September 1939, Turing immediately reported to the Government Code and Cypher School (GC&CS) in Bletchley Park, Milton Keynes. In thorough consultation with the Polish cryptographer (and group theory expert) Marian Rejewski and his team, who in November 1938 had already successfully broken the German cryptosystem "Enigma" with custom-built machinery (the "bomba kryptologiczna"), he quickly started to construct a second apparatus.[508] The aim of the "Bombes" was to break into German communications on a daily basis, and the first one, optimistically named "Victory," was installed in Hut 1 in Bletchley Park on 18 March 1940. Rejewski wrote about the meeting with the man who would become Britain's cryptographic master-mind for the next fourteen years in early 1940: "We treated [Turing] as a younger colleague who had specialised in mathematical logic and was just starting out in cryptology."[509]

The Bombe was an electromechanical, analogue apparatus based on the permutations of the gear-wheeled Enigma rotors, which changed at every step of the encryption procedure. The gears here were not continuous, but discrete, each tooth, as it were, carrying a letter. Like the earlier automatic telephone exchange system, it was a gigantic switching machine, creating and changing connections.

---

507. William Thomson, The tide gauge, tidal harmonic analyser and tide-predicter. *Minutes of Proceedings of the Institution of Civil Engineers* 65 (1881): 2–25.

508. For the *bomba*, cf. Chapter 5.

509. Władysław Kozaczuk, *Enigma: How the German Machine Cipher Was Broken, and How It Was Read by the Allies in World War Two* (Frederick, MD, 1984), p. 97.

The Bombe did not analyse Enigma, it simply integrated 36 of them and in this way exhausted its variance combinatorically.

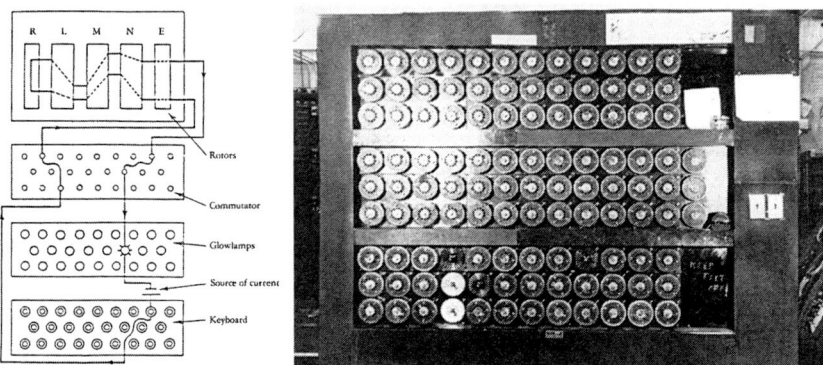

Fig. 4: ENIGMA and Turing Bombe.

One of Turing's main insights was that we cannot build machines that analyse simple properties of other machines in a general way. The Bombe (and earlier, the Polish bomba) showed that it was possible to build an apparatus that could exploit the concrete *weaknesses* of other apparatus, which reverse-engineers, or "hacks" it, after it has been accomplished. Gödel had written in a footnote of his famous paper "On Formally Undecidable Propositions": "The true reason for the incompleteness which attaches to all formal systems of mathematics lies […] in the fact that the formation of higher and higher types can be continued into the transfinite […], while, in every formal system, only countably many are available."[510]

Lorenz SZ40/2 and Colossus

Less than four years later, in January 1944, a much faster machine, Colossus 1, was installed in Block F in Bletchley Park, conceptualised by Max Newman, Turing's former tutor, and developed by Tommy Flowers from the GPO Research Station in Dollis Hill, London, whom Turing had recommended for the job. The target machine in this case was the German Lorenz SZ40/42, a teleprinter incorporating an attachment for encryption used by Hitler to communicate secretly with his generals. These were run in radio or cable networks of twenty-odd participants and worked directly on the 5-bit code used for transmission, by adding a bit stream generated by 12 rotors. These all had different numbers of cams, all of them prime: 23, 26, 29, …, 61, and their period taken together was

510. Kurt Gödel, Über formal unentscheidbare Sätze der *Principia Mathematica* und verwandter Systeme, I. *Monatshefte für Mathematik und Physik* 38 (1931): 173–198, here p. 191, fn. 48a.

bigger than 10^{19}, thus generating a very long non-repeating sequence.[511]

Colossus was a fully-electronic, valve-based machine that executed logical and counting functions on two bitstreams, one coming from the paper tape under consideration, and the other, from an electronic simulation of the stepping rotors of the German SZ40. While running the tape against all possible starting positions of the bitstream coming from the SZ40's rotors, Colossus calculated the *delta* of the sum of the first two bits of each letter from both the message tape and the chi tape and compared the two resulting streams. In case the correspondence was higher than a threshold that could be manually set, the result was printed out on the attached teleprinter. The real starting position of the first two chi-wheels of the SZ40 was expected to yield approximately 55%, slightly more than random, correspondence. Using advanced statistics, the trace of the bit pattern on the rotors could be detected in the ciphertext and hence their starting position could be reconstructed.

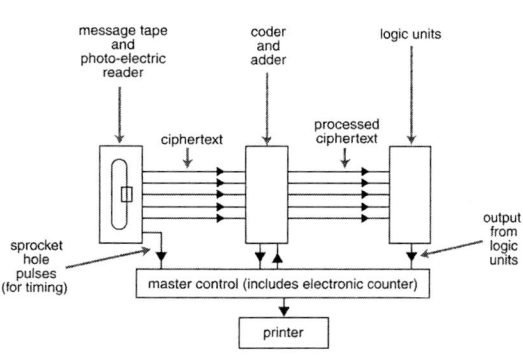

Fig. 5: Lorenz SZ40/42 and Colossus rebuild; schema of Colossus by Flowers.

This method of cryptanalysis had been invented by William Tutte in 1943 on the basis of "Turingery" which also relied on the *delta*, the sum of two temporally-consecutive bits in the stream, and had been devised by Turing in July 1942. It is clear that he must have been informed about all the details of Colossus, although he refused to play an active part in its construction. Also the statistic methods of "Banburism," developed by him and I. J. Good for breaking the naval Enigma, a weight of evidence in the logarithmic scale named *ban*, with sub-units *deci-ban*, played a decisive part in the analysis of "Tunny" traffic, as the Lorenz crypto-teleprinter was called in Bletchley Park. In 1941/2, Turing authored at least three papers: on the cryptanalysis of Enigma, on the statistics of repetitions, and on the applications of probability within cryptography.

---

511. B. Jack Copeland, ed., *Colossus* (Oxford, 2006), pp. 91–100, 352–369.

Again, the enemy machine was integrated into the code-breaking computer, but on a different carrier: Instead of literally incorporating its hardware, like the Bombe and the bomba, Colossus represented the cams on the rotors of the SZ40 by bits on ring counters of thyratron valves. It was the first of Turing's real machines that was completely based on – on/off – bits. The switching on or off of current to bulbs behind the dioptric letter transparencies of Enigma, was abstracted (in the sense of Hegelian *aufheben*) into the simple difference of 0 or 1.

In parallel to Turing's *Computable Numbers*, in 1936, the American mathematician Claude Shannon had recognised that the binary theology of George Boole could be used to calculate networks of relays and that these in turn could solve Boolean algebra problems. He constructed a number of artefacts from these components, the most complex of which was a Factoring Table Machine. Turing first met Shannon in January 1943 when he visited the Bell labs to learn about speech scrambling technology, for his own project in this direction – "Delilah." He also met William Friedman on this occasion, the great American cryptanalyst who had broken the Japanese "Purple" code.

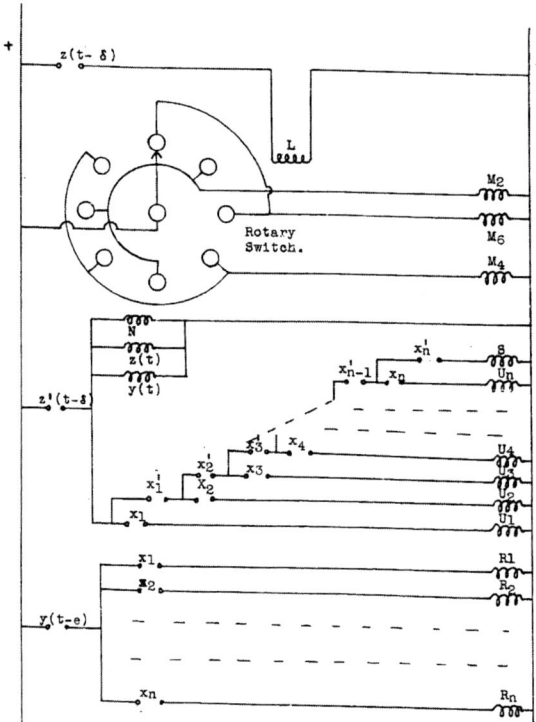

Fig. 6: Shannon's Factoring Table Machine, 1936.

Delilah

Around the same time, from May 1943 until June 1944, Turing constructed and built a speech encryption machine at the secret MI-6 base at Hanslope Park near Bletchley (still active today), together with the electrical engineer Donald Bayley. It merged two technologies employed in his earlier machines, having rotors like the SZ40 on one side and electron tubes like Colossus on the other, and scrambled speech by perfecting the principles of both. Like the Lorenz, it employed the multiplication of numbers with no common denominator like 7, 8, 11, 13, 15 and 17 on the rotors, generating a very long, non-repeating sequence – a length of slightly over 2 million or about eight minutes, as Turing wrote in his report. The output from this key generator was then added in modulo (non-carrying) to a sampling of the speech signal at 4,000 Hz, which allowed a maximum range of 2,000 Hz for the voice input, and then encoded back from pulse into an audio signal via Fourier synthesis.[512]

Contrary to the initial thesis, it seems that Turing's preoccupation from very early on was constructing machines of diverse sorts, probably even more than pure mathematics. The second sentence in Newman's obituary and an implicit verdict about Turing's work from 1945 to 1951 was: "After some years of scientific indecision, since the end of the war, Turing had found, in his chemical theory of growth and form, a theme that gave the fullest scope for his rare combination of abilities, as a *mathematical analyst with a flair for machine computing*" (my emphasis, D.L.).

One thing that could be learned from this genealogy of machines and from the cryptographic speed race of the "battle of wits" was that cryptological apparatus should be constructed as general as possible to be able to be re-used in case the opponent changed his encryption strategy, which again had to be permitted by the universality of his hardware. Cryptology as combinatorics *per se* moved practically towards the universality of Turing's 1936 paper for immediate strategic reasons. The other principal result was that speed was of absolute essence.

Automatic Computing Engine ACE
and Manchester University Computer MUC

Turing's thoughts on building the perfect symbolic Universal Machine became much more concrete in 1945, when he joined the National Physical Laboratory and conceptualised the ACE computer. In "Proposed Electronic Calculator", a report written in early 1946 for the NPL, he surveyed much of the available technology of the time, in particular the different means of electronic storage, and sketched his fifth machine in extensive technical detail.[513]

---

512. A.M. Turing, Delilah report (typescript, 6 June 1944, National Archives, box HW 62/6; transcript by A. Hodges; online: http://www.turing.org.uk/sources/delilah.html).

513. A.M. Turing, Proposed electronic calculator, in: *The Collected Works of A.M. Turing: Mechanical Intelligence*, ed. D.C. Ince (Amsterdam, 1992), pp. 1–86.

Due to administrative hurdles, he left the project in May 1948 for a more promising one in Manchester, the Mark I. Just a few weeks later, on 21 June 1948, the prototype of this machine, the SSEM (Small Scale Experimental Machine), performed its first successful calculation. In October 1948, he moved to Manchester and started to work under the new contract, but already in July, he devised a long division routine (the well-known hand division method) for the prototype, which Tootill wrote down into his notebook, together with an optimised factoring routine he suggested in September. This was again related to the search for prime numbers, like his unrealised plans for the zeta machine.

Fig. 7: Pilot ACE, Science Museum, London.

Date	Artefact	Carriers used	Code
May 1936	Universal machine	paper tape	bits
Jul 1939	Zeta-function machine	brass wheels, pulleys	floating point
Mar 1940	Bombe	rotors, wires, brush contacts, current	letters, on/off
Feb 1944	Colossus	5-bit paper tape, electron tubes, pulses	bits, Baudot 5-bit
Jun 1944	Delilah	rotors, wires, brush contacts, electron tubes, pulses, audio	floating point, sampling, Fourier synthesis
Feb 1946	ACE (report)	5-bit paper tape, electron tubes for switching data, delay lines for storage, pulses	bits, Baudot 5-bit, 32-bit words
Oct 1948	Manchester Mark I	5-bit paper tape, electron tubes for switching data, Williams tubes for storage, pulses	bits, Baudot 5-bit, 20-bit words
Feb 1951	Ferranti Mark I	5-bit paper tape, electron tubes for switching data, Williams tubes for storage, pulses	bits, Baudot 5-bit, 20-bit words

Fig. 8: "Turing's" machines.

Turing's Software

In his new job at the Computing Laboratory of Manchester, according to his biographer Andrew Hodges, Turing specified "the range of operations that should be performed by the machine" in cooperation with the engineers Frederic Williams, Tom Kilburn, Geoff Tootill, and others, who constructed and built the computer. He had no control over its interior logical design, but over the input and output mechanism, for which he proposed to use a teleprinter he subsequently acquired through his GCHQ contacts. Before he arrived, the engineers had already created a switch panel, which would remain the main method of manual input to computers for an astonishingly long time, during which no "keyboard" as it is known today existed.[514]

Turing's first job was to write a user manual and basic programmes for the machine that was by then fully specified and ready to be built by the Ferranti company. From October 1949 to March 1951, together with his two female assistants Audrey Bates and Cicely Popplewell, he developed the first "operating system," Scheme A, which allowed the user to call up one level of subroutines from the main programme. He also created the "formal mode," a debugging software that traced the initial values and the input of the user when testing an algorithm. In March 1951, he released the first edition of the manual.[515]

514. Cf. Chapter 7.

515. A.M. Turing, *Programmers' Handbook for Manchester Electronic Computer Mark II* (typescript, March

In November, he and Geoff Tootill designed a truly-random number generator for the machine based on a noisy diode, which was needed for methods of statistical sampling like Monte-Carlo.

In June 1949, Max Newman and Turing again turned to the investigation of primes, and Newman wrote a software for the Manchester Mark I still in the lab, to search for Mersenne primes, that is, primes of the form $M_p = 2^p - 1$, for p up to 353. Turing proposed an optimised algorithm. In his report from 1949, Newman wrote the programme was the "first routine for a problem with some intrinsic interest to have been run on a 'general purpose' machine."[516] It can therefore be considered the first piece of software ever written, after Ada Lovelace's algorithm for the calculation of the numbers of Bernoulli in 1843, that is, numbers of the form:

$$S_m(n) = \sum_{k=1}^{n} k^m = 1^m + 2^m + \cdots + n^m.$$

The Mersenne algorithm was followed in June 1950 by another one for the investigation of the Riemann zeta function. The conditions were unfavourable when it was tried and much less was achieved than planned: Turing extended the number of zeroes on the critical line from the first 1041 to the first 1104, then the machine broke down.[517]

Around the same time, he developed an encryption algorithm of "only 1000 units of storage" (approximately one CRT screen or half a page of programming, 1280 bits or 64 instructions). When supplied with one sixteen-figure number, it replied with another within two seconds. Turing stressed that it would be impossible to derive its response to a new input from its past replies. This software is probably still classified, or lost. It is a curious fact that Alan Turing, the person who made programming possible by inventing the concept of the Universal Machine and then helped it to be built, did not write many longer algorithms when it was possible to do so for the last five and a half years before his death.

May 1936	On computable numbers
March 1939	A method for the calculation of the zeta-function
July 1948	Long division routine
June 1949	Mersenne primes w. Newman
June 1950	Some calculations of the Riemann zeta-function
June 1950	Cipher programme
March 1951	Scheme A, formal mode
May 1951 – June 1954	Morphogenesis

Fig. 9: Turing's programmes.

1951; online: http://www.turingarchive.org/browse.php/B/32).

516. M.H.A. Newman, Some routines involving large integers, in: *Report of a Conference on High Speed Automatic Calculating Machines, 22–25 June 1949*, ed. Maurice V. Wilkes (Cambridge, 1950), pp. 73–75.

517. A.M. Turing, Some calculations of the Riemann zeta-function. *Proceedings of the London Mathematical Society (3) 3* (1953): 99–117.

The Art of Programming

Already before it was replaced by the industrially-produced Ferranti Mark I in February 1951, some interesting scientific software developments had taken place on the Manchester Mark I prototype. An early programme was Audrey Bates' thesis, *On the Mechanical Solution of a Problem in Church's Lambda Calculus* from October 1950.[518] Even more momentous was the Monte-Carlo analysis pioneered here already in summer 1950 by Stan Frankel and Berni Alder from California, USA; the former had also been part of Oppenheimer's core group at the Manhattan Project in Los Alamos. Due to a lack of computing resources in North America at the time, Frankel travelled to Manchester to run Alder's thesis project on the Mark I. Since the latter's advisor was not convinced of the investigation method of Monte-Carlo sampling, the results were only published much later, in 1955.[519]

In October 1951, Tony Brooker was hired by Turing to take over his role as a system developer and administrator, thereby completely freeing him from any formal job descriptions. In November 1951, Christopher Strachey, who had first visited the lab in July, joined in as an employee of the National Research and Development Corporation NRDC (a contract which officially started in June 1952), to conduct "a serious long-term study of the art of programming," and in Andrew Hodges words, "Alan's days as grand master of the console were over. He had handed on the torch" to the man who would become the first full-time software developer in the UK.

As mentioned earlier, the former schoolteacher Strachey had written a very long algorithm for playing the game of draughts on the Pilot ACE in February 1951, which completely exhausted its memory.[520] When he heard about the Manchester Mark I, which had a much bigger storage, he asked his former fellow-student Alan Turing for the manual, translated his software into the operation codes of that machine by around July 1951, and was given permission to run it on the computer. During the summer holidays, at the suggestion of Turing, he developed another algorithm of amazing length, a debugging programme named "Checksheet" of twenty-odd pages.[521]

Surprisingly, it is quite difficult to find out what exactly this first software was, since there are several versions of the above story in the literature based on oral histories. Most likely, Strachey was demonstrating his draughts algorithm in October 1951, and since he was still developing it, it was running in the "Checksheet" debugging environment he had created. This programme played

518. Audrey Bates, *On the Mechanical Solution of a Problem in Church's Lambda Calculus* (M.Sc. thesis, Manchester University, 1950).

519. Bernie J. Alder, Stan P. Frankel, and Victor A. Lewinson, Radial distribution function calculated by the Monte-Carlo method for a hard sphere fluid. *The Journal of Chemical Physics* 23 (1955): 417–419.

520. Cf. Chapter 7.

521. Martin Campbell-Kelly, Programming the Mark I. Early programming activity at the University of Manchester. *Annals of the History of Computing* 2 (1980): 130–168; M. Campbell-Kelly, Christopher Strachey, 1916–1975. A biographical note. *Annals of the History of Computing* 7 (1985): 19–42.

"God Save the King" (it was written and executed in 1951!) when a debugging session was exited, and this is why the people attending were under the impression that the draughts software played the national anthem at the end. The author has located the "Checksheet" algorithm including the tune-playing part (one of the first musical computer programmes ever) in Bodleian library and this will be the topic of a separate investigation.[522]

In 1952, Christopher Strachey also wrote a software for the generation of love-letters, which shows the very early interest in Artificial Intelligence in Manchester.[523]

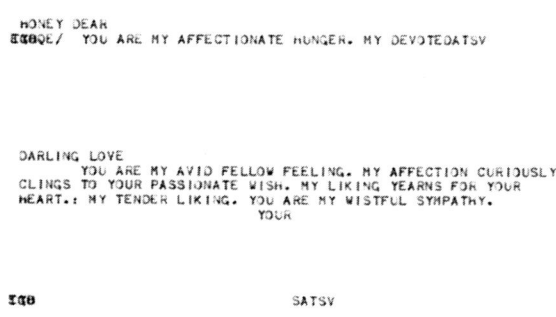

Fig. 10: Artificial Intelligence: Love letters algorithm.

From May 1951 to June 1954, Turing worked on a theory and software of morphogenesis, of which only amazingly small programming fragments survive.[524] This may be due to the fact that he was a master of "optimum coding." In summer 1953, Bernard Richards arrived in the Manchester laboratory and started a PhD under his guidance, "Morphogenesis of Radiolaria."[525]

522. National Cataloguing Unit for the Archives of Contemporary Scientists (NCUACS), Catalogue of the Papers and Correspondence of Christopher Strachey (1916-1975) (CSAC no. 71/1/80; online: http://discovery. nationalarchives.gov.uk/browse/r/24dc1102-e60d-489e-9443-55b0af464f1a).

523. Cf. Chapter 3.

524. A.M. Turing, The chemical basis of morphogenesis. *Philosophical Transactions of the Royal Society of London Series B* 237 (1952): 37–72.

525. Bernard Richards, *Morphogenesis of Radiolaria* (M.Sc. thesis, Manchester University, 1954).

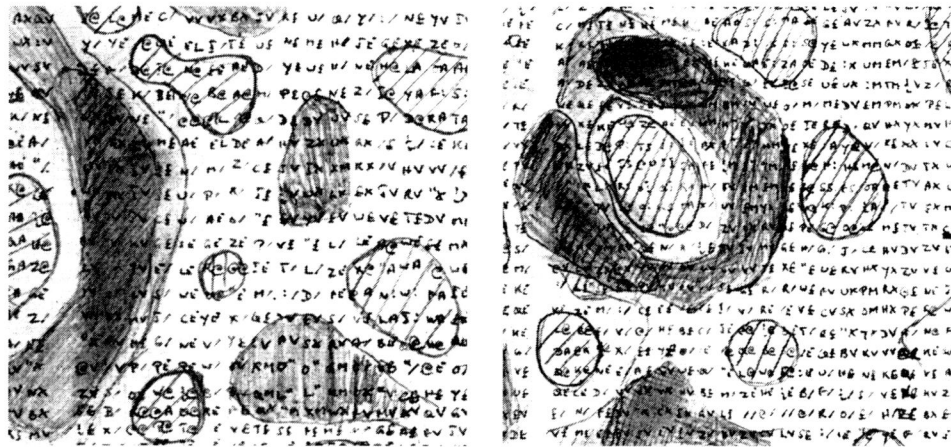

Fig. 11: Turing's morphogenesis project.

Turing also helped other people with their mathematical problems, like the engineer Robert K. Livesley, who was developing algorithms to calculate minimum-weight structures. Engineering was one of the main disciplines in the early computerisation of science. Strachey's first official programming duties were also in that area. In the "Garry" project for Halcrow & Partners in January 1952, he designed a surge shaft for the water power station at the Quoich dam in Scotland. In September/October 1952, he worked on the St. Lawrence Seaway and Power Project on FERUT, the Canadian counterpart of the Manchester machine at the University of Toronto. The algorithm calculated the backwater flow in the St. Lawrence Seaway between Prescott and Cornwall, Ontario, under the changed conditions of the deepening of the navigational passages and the construction of a power dam at Cornwall, by assessing the change in water levels and surface profiles.

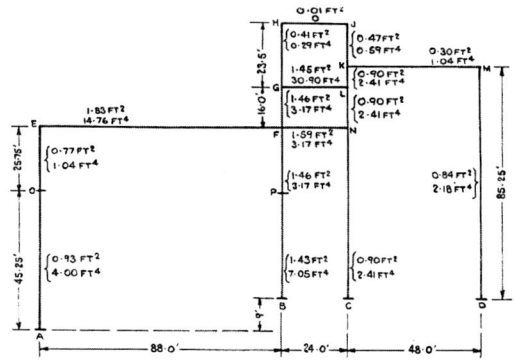

Fig. 12: Robert K. Livesley, Minimum-weight structure for power-station building, 1954.

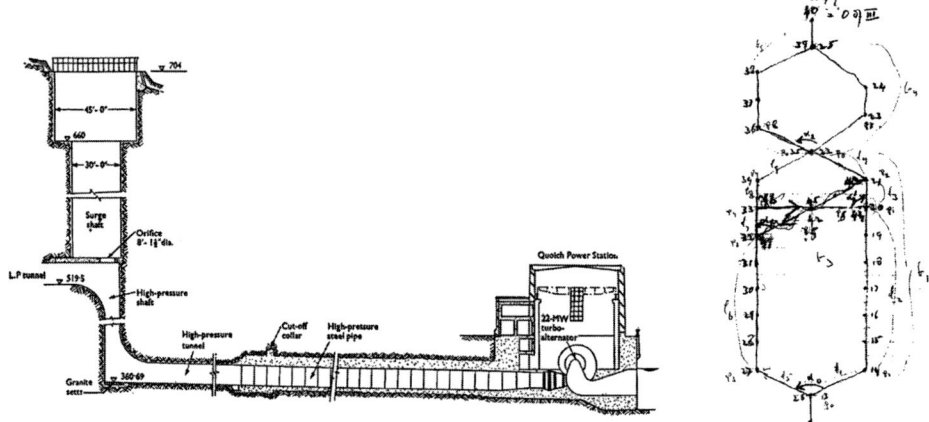

Fig. 13: Profile of Quoich stage; a part of Strachey's formalisation of the St. Lawrence Seaway.

Other important scientific disciplines in the early period of programming are X-ray crystallography, nuclear physics, quantum chemistry and meteorology.

The core of the computerisation of British X-ray crystallography consists in the software developed by Farid Ahmed in the years 1952/53, which comprises over 100 pages.[526] The eminent crystallographer Durward Cruickshank and others employed it in several research projects. It was the standard software for the task at the time. The British scientist Dorothy Hodgkin (1910–1994), who won the Nobel price in chemistry in 1964, also performed calculations on the Mark I, and her application of X-ray technology to biologically interesting molecules led to the rise of genetics only shortly later.

It is well-known that the military and civil use of nuclear energy played an important role in early computer usage. At the Mark I there was one user who worked for the "Atomic Weapons Research Establishment" (AWRE) in Aldermaston, whose investigations were so secret that he had to replace the ink ribbon of the teleprinter at the end of each shift. Alick E. Glennie, who in 1952 also wrote the first proto-compiler, "Autocode," for the Mark I, performed Monte-Carlo simulations of nuclear processes and their chain reactions on the machine.

526. Farid R. Ahmed, *Development of Mathematical Methods for the Determination of Molecular Structures by X-ray Analysis* (Ph.D. thesis, Leeds University, 1953).

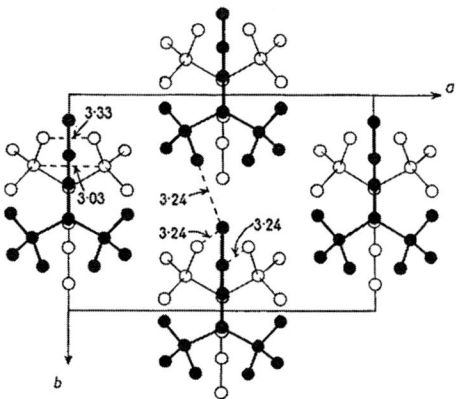

Fig. 14: Computational problems in X-ray crystallography.

In quantum chemistry, Huw Owen Pritchard and Frank Sumner authored a number of programmes from 1954 to 1956. One of them calculates a cross-section through the hydrogen-atom wave function, another determines Hückel bond orders and polarisabilities and the third calculates the secular matrix for the LCAO description of the H_2^+ molecule-ion.

The computerisation of British meteorology was pursued by Fred Bushby and Mavis Hinds, who computed early weather prediction models on the Mark I. From 1954 on, 24-hour forecasts based on the Sawyer–Bushby two-parameter atmospheric model equations with a resolution of one hour were produced within about 4 hours computing time on the Mark 1.

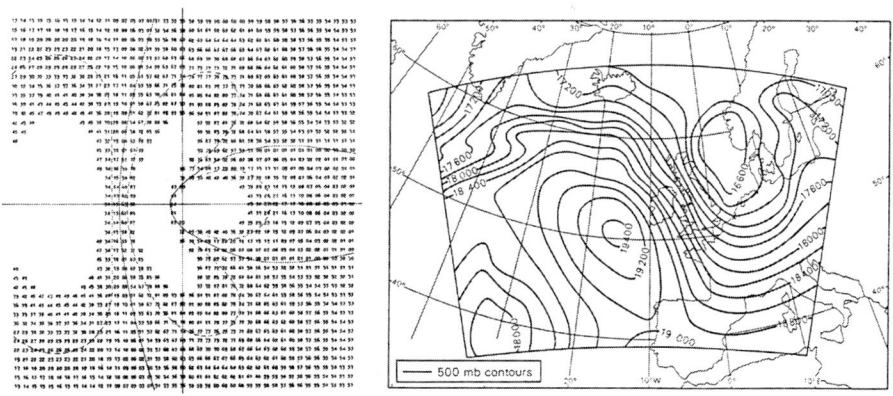

Fig 15: Applications in quantum chemistry and meteorology.

May 1951 – Jul 1952	Draughts	Christopher Strachey
Aug/Sep 1951	Checksheet, God Save the King	C. Strachey
Jan 1952	"Garry" surge shaft	C. Strachey
Sep–Oct 1952	St. Lawrence Seaway	C. Strachey
1952–3	X-ray crystallography	Farid Ahmed, Durward Cruickshank
1953	Love-letters	C. Strachey
1953–4	Minimum-weight design	Robert Livesley
1953–4	Monte-Carlo simulations, nuclear physics	Alick Glennie
1953–4	Morphogenesis of Radiolaria	Bernard Richards
1954–8	Meteorology	Fred Bushby, Mavis Hinds
1954–6	Quantum chemistry	Huw Pritchard, Frank Sumner

Fig. 16: Overview of software developments on the Ferranti Mark I, 1951–1958.

Towards an Archaeology of Algorithmic Artefacts

Up until its dismantling in 1958, numerous scientists employed this first fully-electronic computers in their research. Their usages of the machine are essential in the sense that they would not have developed their theories as fast and in the same way without the machinic support of their calculations. The scientists made the results of their computerised research available to their colleagues via articles, but the programmes that produced them circulated informally at most and were often lost when a new generation of computers came along. Just a few years later, the obsolete machines themselves were taken down and disassembled into their components, which were often re-used in other projects.

Even if the source code of a certain calculation for the Mark I can be located, the next problem lies in the fact that at the time, the programmes consisted in operation codes using the Baudot alphabet, which cannot be understood by simply reading them. The only possibility for investigating algorithms of a certain complexity is to re-run them and to study their behaviour.[527] It is for this reason that I created the "Write Current" project with the aim to not only situate and investigate the totality of Mark I software between 1948 and 1958, but also to execute it on an emulator of the Ferranti Mark I (a simulation of the machine) in order to gain a concrete understanding of its functioning. This process of reproducing algorithms in order to study them, that is, my attempt at an Archeology of Algorithmic Artefacts, has

---

527. Cf. Chapter 4.

already been achieved for Strachey's love-letters and his draughts programme, and the source code of most of the software tabulated in Figure 16 has been safeguarded.

Of the impressive list of Turing's machines tabulated above, not a single one has survived, even though the earliest one that was successfuly produced, the Bombe, was built only 75 years ago, with slightly over 200 copies having been made. The number of surviving witnesses from these times is rapidly dwindling. This makes it extremely difficult in the important formative period of these disciplines to comprehend not only the solutions, but also the problems and approaches, as is postulated in the formula of the reproducibility of experiments for all the sciences. As for Turing's algorithms (cf. the figure), almost nothing survives.

Bibliographic References

Chapter 1: while(true). On the Fluidity of Signs in Hegel, Gödel, and Turing, was first published in: Variantology 1. On Deep Time Relations of Arts, Sciences and Technologies, eds. Siegfried Zielinski and Silvia Wagnermaier (Cologne: König, 2005), pp. 261–278. Translated by Gloria Custance.

Chapter 2: Traces of the Mouth. Andrei Andreyevich Markov's Mathematisation of Writing, was first published in: History of Science 44.145 (2006): 321–348. Translated by Gloria Custance and David Link.

Chapter 3: There Must Be an Angel. On the Beginnings of the Arithmetics of Rays, was first published in: Variantology 2. On Deep Time Relations of Arts, Sciences and Technologies, eds. Siegfried Zielinski and David Link (Cologne: König, 2006), pp. 15–42.

Chapter 4: Enigma Rebus. Prolegomena to an Archaeology of Algorithmic Artefacts, was first published in: Variantology 5. Neapolitan Affairs, eds. Siegfried Zielinski and Eckhard Fürlus (Cologne: König, 2011), pp. 311–345, and in: Interdisciplinary Science Review 36.1 (2011): 3–23. Copyright © Institute of Materials, Minerals and Mining, reprinted by permission of Taylor & Francis Ltd, www.tandfonline.com, on behalf of Institute of Materials, Minerals and Mining.

Chapter 5: Resurrecting Bomba Kryptologiczna: A Reconstruction of the Polish Crypto Device, was first published in: Cryptologia 33.2 (2009): 166–182. Copyright © Taylor & Francis Group, LLC. The author is indebted to Anna Ucher for translating relevant parts of Rejewski's accounts from Polish.

Chapter 6: Scrambling T-R-U-T-H. Rotating Letters as a Form of Thought, was first published in: Variantology 4. On Deep Time Relations of Arts, Sciences and Technologies in the Arabic–Islamic World and Beyond, eds. Siegfried Zielinski and Eckhard Fürlus (Cologne: König, 2010), pp. 215–266.

Chapter 7: Programming ENTER. Christopher Strachey's Draughts Programme, was first published in: Resurrection. The Journal of the Computer Conservation Society 60 (2012/3): 23–31.

Picture Credits

Jason Wagner, Drew S. Burk
(Editors)
Univocal Publishing
411 Washington Ave. N., STE 10
Minneapolis, MN 55401
www.univocalpublishing.com

ISBN 9781937561048
This work was composed in Times New Roman.
All materials were printed and bound
in May 2016 at Univocal's atelier
in Minneapolis, USA.

The paper is Hammermill 98.
The letterpress cover was printed
on Lettra Pearl.
Both are archival quality and acid-free